To Zak

D1282996

With best wishes
from
Liz Strachan

Liz Strachan was a maths teacher for thirty-six years. Her writing career started just before she retired when, on a whim, she entered and won first prize in the European Letter Writer of the Year Competition. Since then, she has published about 200 articles and short stories and, at the Scottish Association of Writers' Annual Conference, she has won six first prizes and numerous other awards. She is married, and has two sons and four grandchildren.

Numbers are Forever

LIZ STRACHAN

Constable • London

Constable & Robinson Ltd
55-56 Russell Square
London WC1B 4HP
www.constablerobinson.com

First published in the UK by Constable,
an imprint of Constable & Robinson Ltd., 2014

A copy of the British Library Cataloguing in Publication
Data is available from the British Library

ISBN 978-1-47211-104-3 (paperback)
ISBN 978-1-47211-110-4 (ebook)

Printed and bound by CPI Group (UK) Ltd, Croydon, CR0 4YY

1 3 5 7 9 10 8 6 4 2

Contents

Contents

Contents

Acknowledgement

I wish to thank my editor, Hugh Barker (who admits to a certain fondness for the subject), for suggesting that I should write *Numbers are Forever* and encouraging and supporting me throughout.

Introduction

I do not claim to be a brilliant mathematician. As George Bernard Shaw said, 'He who can, does. He who cannot, teaches.'

However, I was an enthusiastic teacher of mathematics for thirty-six years and, more importantly, I was a teacher of teenagers, many of whom were not avidly keen on the subject and needed to be cajoled, diverted and entertained. Maths teachers particularly hate the dreaded last twenty minutes on a Friday afternoon when teenage brains have shut down for the weekend and the curriculum subject for the week is trigonometric identities. That is the time to forget about the heavy stuff, the quadratic equations and the compound interest and indulge, just for a short while, in the magic of numbers.

This book is only about numbers – that is, whole numbers – whole numbers and nothing but whole numbers. The whole numbers start 0, 1, 2, 3, 4 … and go on and on forever (see the Glossary for the distinction between whole numbers, natural numbers and integers). Mathematicians love them and the nineteenth-century German scholar Leopold Kronecker was so enamoured that he wrote, 'God himself made the whole numbers. Everything else is the work of man.'

Negative numbers, mixed numbers, complex numbers, rational numbers, irrational numbers, real numbers and – God forbid! – imaginary numbers will never be mentioned. There will be no algebra, geometry or trigonometry. Fractions and decimals will also be taboo – well, most of the time.

The Interesting Numbers Paradox states that there are no uninteresting whole numbers. If one mathematician remarks that, for example, 74 is a rather dull number, another mathematician will immediately be very interested to find out why and therefore 74 becomes an interesting number. (And it is, as you will find out later!) Having proved in a tongue-in-cheek way that all numbers are interesting, there are, of course, some numbers that are more interesting than others and these are the ones you will see in this book.

So, *Numbers are Forever* is not for mathematical geniuses or fanatical number nerds but it is for everyone to enjoy, whatever their age or education.

The Very Beginning Starts with Zero

Once upon a time, there were no zeros.

The Greeks, who gave us Euclidean geometry and the Theorem of Pythagoras, had no concept that nothing or emptiness could be expressed as a number. Numbers, as far as they were concerned, started at 1.

The Romans had no need for a 'nothing' symbol either. They had the letters M, D, C, L, X, V and I. So, 1003 was written MIII and 365 was written CCCLXV.

The next big advance (and it was a huge advance) came in the sixth century in India. Mathematicians there created a different symbol for every number from 1 to 9, which curiously became known as Arabic numbers. And then they created an entirely new number for 'nothing', which was later called zero.

Once zero was invented, it transformed counting in a way that would change the world. The concept of nothing or emptiness now had a number.

The zero, on its own, wasn't necessarily all that special. The magic happened when it was paired with other numbers to make them larger or smaller. It also made calculation much easier.

By the twelfth century, Arabic numbers had found their way to North Africa. From there, they were introduced into

Europe, thanks to the brilliant son of the chief magistrate from Pisa who administered Italian trading in Algeria.

Leonardo of Pisa (later in this book called by his other name, Fibonacci), who previously had used Roman numerals back home in Italy, was enthralled with this new number system. On his return in 1202, he produced a book called *Liber Abaci* (*The Book of Counting*) that extolled the Arabic number system and, slowly but surely, the numbers, led by zero, spread throughout Europe. The rest is history.

In maths, the symbol 0 is always called zero. We don't call it 'nothing', 'nought', 'oh', 'nil' or 'zilch'. The zero is vitally important for keeping the other numbers in the correct place.

Early in school, we learned the headings:

Thousands	Hundreds	Tens	Units
1	0	0	3

The difference between 13 and 1003 is only a matter of 2 zeros but they are very important zeros because they keep the digits in their proper place.

When zeros are at the end of a number, it is important to get them correct.

For instance, a yellowing piece of paper was found in the desk of deceased Aunt Beatrice, on which she had written, 'I, Beatrice Mills, being of sound mind, do hereby bequeath to my dearest niece and nephew the following sums of money: Victoria Mills £500000, Hugh Mills £50000.'

Now, did Aunt Beatrice accidentally miss out a zero in Hugh's bequest? Or had she never forgiven him for forgetting her 80th birthday?

To avoid confusion, it is customary to add commas to numbers with 5 (sometimes 4) or more digits. Aunt Beatrice should have grouped the zeros in batches of 3, starting from the right – £500,000 or £50,000. As it was, Hugh got £450,000 less than his sister.

In this book, however, commas will occasionally be omitted, even in very large numbers, in order to emphasize number patterns.

Adding, Subtracting, Multiplying and Dividing by Zero

$2 + 0 = 2$
$2 - 0 = 2$
$2 \times 0 = 0$
$2 \div 0 = ?$

Something seems terribly wrong here. The calculator shouts 'ERROR'. This is because the little machine just can't do this calculation. Nor can a super computer.

A number divided by a large number gives a small answer. A number divided by a small number gives a large answer. So if a number is divided by a very small number like 0.000001, you get a very large answer. A number divided by the smallest number of all, which is 0, will give the largest possible answer – but of course there is no 'largest number' as numbers go on forever. This is where maths breaks down because 2, or any other number, just cannot be divided by 0. By convention, we use 'infinity' as the answer but 'infinity' is not the very largest number because no such number exists. Some mathematicians prefer to take the easy way out and say that any number divided by zero is undefined.

Albert Einstein, the most influential physicist of the twentieth century, was the first to say that black holes were a result of God dividing the universe by zero.

0^0

What about 0^0? Well, that's easy, isn't it? In the Glossary, it states that any number to the power of 0 is 1. Ah, but with one exception, and that exception is 0^0. Definitely correct are: $1^1 = 1$, $1^0 = 1$, $0^1 = 0$. But what *is* the correct answer for 0^0? Again, mathematicians prefer to say that 0^0 is undefined but that is the same as saying 'nobody knows'.

Natural Numbers

The natural numbers originate in the words used from time immemorial to count things like 2 dinosaurs or 3 hairy mammoths or perhaps 1 fat dodo for the family dinner. Zero, at that time, was never needed. So the natural numbers are: 1, 2, 3, 4 … and they go on for ever.

To demonstrate straight away just how amazing numbers are, write out a string of them:

1, 2, 3, 4, 5, 6, 7, 8, 9, 10, 11, 12, 13, 14, 15, 16, 17, 18, 19, 20, 21, 22, 23, 24, 25, 26, 27, 28, 29, 30 …

Now arrange them into groups of 1 digit, then 2 digits, 3 digits and so on:

1   ~~2, 3~~   4, 5, 6   ~~7, 8, 9, 10~~   11, 12, 13, 14, 15
~~16, 17, 18, 19, 20, 21~~   22, 23, 24, 25, 26, 27, 28 …

and immediately cross out the 2nd, 4th, 6th and all other even-numbered groups. So you are left with:

1   4, 5, 6   11, 12, 13, 14, 15   22, 23, 24, 25, 26, 27, 28
…

Now add the groups as follows:

$$\text{Sum of one group} = 1$$
$$= 1^4$$
(all will become clearer in a moment!)
$$\text{Sum of 2 groups} = 1 + 4 + 5 + 6$$
$$= 16$$
$$= 2 \times 2 \times 2 \times 2$$
$$= 2^4$$
$$\text{Sum of 3 groups} = (16) + 11 + 12 + 13 + 14 + 15$$
$$= 81$$
$$= 3 \times 3 \times 3 \times 3$$
$$= 3^4$$
$$\text{Sum of 4 groups} = (81) + 22 + 23 + 24 + 25 + 26 + 27 + 28$$
$$= 256$$
$$= 4 \times 4 \times 4 \times 4$$
$$= 4^4$$

Now try out 5 groups for yourself (and any more if you are really keen) and if you don't get 5^4 and 6^4, your adding is disgraceful!

Nicomachus

There is no person more annoying than a numbers geek who holds on to you like the Ancient Mariner who waylaid the Wedding Guest and wouldn't let him go until he had told his tale about shooting the albatross. However, the numbers geek starts his tale by saying, 'Think of a number between 1 and 100 but don't tell me what it is ...'

The mathematician Nicomachus of Gerasa (now Jerash in Jordan) lived in the first century AD. Although Pythagoras was around a few centuries earlier, Nicomachus was an ardent Pythagorean and took his subject equally seriously. He liked to share his knowledge with his fellow citizens.

Although he was a numbers fanatic, Nicomachus was *their* numbers fanatic. The locals were always happy to cooperate with their eccentric pet mathematician when he said, 'Think of a number between 1 and 100 ...' so just this once, we'll go along with it too.

Having chosen your secret number, say 66, divide it by 3 and say only what the remainder is. Then divide 66 by 5 and state what the remainder is. Once again, do the same for 7. So, you have 3 remainders, 0, 1 and 3. Nicomachus multiplied the 0 by 70, the 1 by 21 and the 3 by 15, getting $0 + 21 + 45 = 66$.

However, if you had chosen, say, 37, your remainders would have been 1, 2 and 2, and $(1 \times 70) + (2 \times 21) + (2 \times 15) = 142$, which is obviously not between 1 and 100. However, that didn't bother Nicomachus. He merely subtracted 105 and got 37.

Sometimes, if a number like 89 is chosen, and the remainders are 2, 4 and 5, giving $(2 \times 70) + (4 \times 21) + (5 \times 15) = 299$, subtracting 105 wasn't enough, but that was no problem,

Nicomachus just subtracted another 105 getting 299 − 105 − 105 = (thank goodness!) 89.

His numeracy skill was prodigious but his reputation for mind-reading powers was quite undeserved.

No Need for Zero

All these sums use only the digits 1 to 9, once each:

$$243 + 675 = 918$$
$$341 + 586 = 927$$
$$154 + 782 = 936$$
$$317 + 628 = 945$$
$$216 + 738 = 954$$
$$215 + 748 = 963$$
$$318 + 654 = 972$$
$$235 + 746 = 981$$

The 8 totals are consecutive multiples of 9.

Prime Numbers

This is a book about whole numbers and nothing but whole numbers. So why introduce prime numbers now? Don't panic! Prime numbers are a special subsection of whole numbers and, like the parent company, they go on for ever.

A prime number is a number that is only divisible by itself and 1. The mathematics police won't allow 1 to be a prime number so 2 is the first prime number and the only one to be even. It is followed by 3, 5 and 7.

	2	3		5	6	7			
11	12	13				17	18	19	
		23	24					29	30
31					36	37			
41	42	43				47	48		
		53	54					59	60
61					66	67			
71	72	73					78	79	
		83	84					89	90
						96	97		

In the table, we have not inserted any more even numbers or numbers divisible by 3 or 5 or 7 and we are left with the 25 prime numbers from 1 to 100.

This process is called the Sieve of Eratosthenes. This great third-century BC mathematician, poet, athlete, astronomer and geographer, despite all his achievements, was nicknamed 'Beta', the second letter in the Greek alphabet. The 'Alpha' mathematician, of course, was his older friend and mentor, Archimedes, who after 23 centuries is still considered to be one of the greatest mathematicians of all time.

Now, in italics, we have inserted all the 6 times table numbers. Sit back and study the table. You can see that every bold prime number (except 2 and 3) is sitting before or after an italic multiple of 6.

You have discovered that all prime numbers (excluding 2 and 3) are of the form:

$$6n \pm 1$$

That is, a number in the 6 times table with 1 added or 1 subtracted.

If only it were that simple! Suppose 91 had been left in the above table by mistake, the $6n \pm 1$ formula would have still worked but 91 divides by 7 so it is not prime. So all prime numbers are of the form $6n \pm 1$, but not all numbers of this form are necessarily prime numbers. It's like saying, 'All strawberries are red berries but red berries are not necessarily strawberries.'

Remember, there are infinitely more prime numbers beyond 100. There are another 143 of them between 100 and 1000 for a start, and beyond that they go on and on forever.

So, for a quick test of whether a number might be prime, first of all write it down with the number before it and after it and test for divisibility by 6. For example, is 981 prime?

980 **981** 982

Neither 980 nor 982 divide by 6 so 981 is definitely *not* prime. In fact, 3 is an obvious factor.

817 has a prime number look about it so let's see:

816 **817** 818

816 divides by 6 but does that mean that 817 is *definitely* prime? Unfortunately, no. All we can say is that it *might* be prime. A closer inspection of the sneaky number reveals that 19 and 43 are divisors so 817 is not prime.

What about 743?

742 **743** 744

742 does not divide by 6 but 744 does. But 743 may be another cunning pretender like 817. However, a laborious lot of dividing by 3, 7, 11, 13, 17, 19, 23 and 29 shows that it is a bona fide prime. (No need to go any further than necessary! 29^2 is bigger than 743 so any number with a divisor higher than 29 will already have been found.)

The formulae $4n + 1$ and $4n + 3$ also produce prime numbers that go on and on but both have failures very early: $4 \times 1 + 1 = 5$ (prime) but $4 \times 2 + 1 = 9$ (not prime); $4 \times 1 + 3 = 7$, $4 \times 2 + 3 = 11$ (prime) but $4 \times 3 + 3 = 15$ (not prime). These formulae go on and on producing primes but they are also riddled with failures.

However, in a letter dated 25 December 1640 to his friend and fellow mathematician, Marin Mersenne, Pierre de Fermat stated that the primes that *are* produced by 4n + 1, when n = 1, 3, 4, 7, 9. 10, 13, 15 ... can all be written as the sum of 2 squares. For example, when n = 15, 4n + 1 = 61 = $5^2 + 6^2$ and when n = 18, 4n + 1 = 73 = $3^2 + 8^2$. This neat bit of maths became known as Fermat's Christmas Theorem. Like his Last Theorem, Fermat probably never proved it but it was proved later by several mathematicians including Leonhard Euler, one of the greatest mathematical geniuses of all time.

There are endless other simple formulae that can create countless prime numbers. For example, 8n + 7 provides 7, 23, 31 and 47 but it fails when n = 6 because 8 × 6 + 7 = 55, which is not prime.

The great German early nineteenth-century mathematician with the fine-sounding name, Johann Peter Gustav Lejeune Dirichlet, proved in the theorem named after him that the above simple formulae and other similar ones go on producing prime numbers ad infinitum although, of course, they do not work for every value of n along the number line. However, the proof is so challenging that in their classic 1938 text, *An Introduction to the Theory of Numbers*, the usually explicit mathematicians G. H. Hardy and E. M. Wright wrote 'This theorem is too difficult for insertion in this book.' Well, if it is too demanding for them, it is impossible for mere mortals. So all we need to know is that, even within arithmetic progressions like 8n+7 or 4n + 3, prime numbers definitely go on and on for ever and ever. OK?

Other prime-rich formulae will be discussed later but, in the end, they are all equally disappointing. It isn't as if mathematicians haven't been trying. They have been obsessed with primes for centuries. Marin Mersenne, who will get a special mention later on when we look at 'The Mersenne Numbers', discovered a formula that was particularly productive. He spent a lifetime studying enormous numbers with only the help of a quill.

Prime numbers and their properties were first studied by the Ancient Greek mathematicians in 500 BC and have been ever since. They remain interesting because no one has ever really cracked prime numbers. In spite of the enormous brainpower of the world's most eminent mathematicians, no foolproof formula exists to create new primes or to check which numbers definitely *are* prime. The work of checking goes on but, nowadays, the hard slog of looking for divisors is done by computers. The prime numbers go on and on along the number line. There will always be more prime numbers. All prime numbers, apart from 2 and 5 at the very beginning, end with 1, 3, 7 or 9 and they are in almost equal numbers. In the Sieve of Eratosthenes showing primes from 1 to 100 on page 10 (and excluding 2 and 5):

5 end with 1
7 end with 3
6 end with 7
5 end with 9

Examining the numbers from 1 to 1000 reveals that there are 168 prime numbers and again, omitting 2 and 5, the frequency of the last digit is

40 end with 1
42 end with 3
46 end with 7
38 end with 9

A study of millions of primes shows that the last digits are random but almost equal in number.

Every so often – in the perennial Math-Olympics – another record-breaking mega-multi-digit prime number is discovered and mathematicians are beside themselves with excitement and determined to find the next one.

Schoolchildren often ask their maths teacher, 'What use is all this stuff?' The answer is that often the most apparently useless of mathematical studies can turn out to be of great importance in the end. In particular, Gottfried Wilhelm Leibniz (1646 –1716) invented binary numbers. In his wildest imaginings, he could not have foreseen that his base 2 numbers would be used in machines called computers found in almost every home throughout the civilized world.

And now, after many centuries of being studied for their own sake, prime numbers are used in encryption software and technology.

Take Any Prime Number

Take any prime number greater than 3, square it, add 11 and divide by 12. It should work out exactly without a remainder. Examples:

$(23^2 + 11) \div 12 = 45$
$(97^2 + 11) \div 12 = 785$
$(823^2 + 11) \div 12 = 56,445$
$(200597^2 + 11) \div 12 = 3,353,263,035$

Try it for any prime number you like. The formula works for all of them, except 2 and 3. Could it be that easy? Could it be that elusive formula for confirming that a number is prime? Sadly, no.

It works for ALL odd numbers except those divisible by 3.

Mathematical joke: 2 is the oddest prime number of all because it is the only one that is even.

A Reminder

If you glance at these 2 simple calculations, $8 + 2 \times 3$ and $21 - 14 \div 7$, and think the answers are 30 and 1, you need a gentle reminder. Even with numbers, there is a pecking order and multiply and divide must always be done before add and subtract. So the above answers are 14 and 19. Likewise, $18 - 8 \times 3 \div 12 = 16$.

The only exception happens when there are brackets. If we add brackets to the calculations above, $(8 + 2) \times 3 = 30$ and $(21 - 14) \div 7 = 1$.

This can be summarized in the rule of BODMAS. **B** is for brackets. Brackets always come first. **O** is for 'of' and it comes next in order of importance (for instance 8 − ½ of 6 = 5). Dividing and Multiplying come next. Adding and Subtracting come last in order of importance, and are of equal status so they can be used in any order.

Goldbach's Conjecture

A theorem is a big mathematical idea that has been proved beyond all doubt. A conjecture is a big mathematical idea that has not been proved to the satisfaction of mathematicians. It seems that every mathematician of any importance, if they can't actually achieve theorem status, like Pythagoras, wants to have their own conjecture.

Christian Goldbach, the eighteenth-century German mathematician, made many important advances in the subject but it is his conjecture that has interested and teased scholars throughout the years. It is a simple idea, easy to demonstrate, but its proof has remained a mystery to this day. He asked, 'Is every even number bigger than 2 the sum of 2 primes?'

He wrote to his good friend and fellow mathematician, Leonhard Euler, but for once, he was also stumped. It certainly seems to be true:

$4 = 2 + 2$
$6 = 3 + 3$
$8 = 3 + 5$
$10 = 3 + 7$

$18 = 7 + 11$
$20 = 3 + 17$

100 = 3 + 97
1000 = 3 + 997
1850 = 883 + 967

And on and on.

No one has ever found a glitch, but until someone comes up with the ultimate proof, Goldbach's discovery remains a conjecture only.

1

The Latin word for one is *'unus'* from which the English words 'unit, unity, unique' and others are derived. Some wonderful clichés are all about 'one': 'One foot in the grave'; 'One for the road'; 'One swallow doesn't make a summer'; 'A one-horse race.'

In maths, 1 is used more than any other number.

Pythagoras (he of the famous theorem) preached to his followers that numbers ruled the universe. The number 1 was especially important. He believed that nothing can exist without a centre, so he started with a point and drew a circle round it. This symbol was called a monad and for Pythagoras, it represented the number 1. He called 1 The First, The Essence, The Foundation, Unity. 1, he thought, was God, the origin of all things.

But, as you learned in a previous chapter, 1 is not the first prime number. It is not the origin of all these wonderful numbers that have fascinated mathematicians for centuries. This is because a prime number by definition has 2 divisors, itself and 1. Poor 1 has only itself as a divisor and therefore is not allowed to be prime. However, to compensate, 1 is the first number in dozens of other number series, as you will see later.

The Pirahã people are an isolated hunter-gatherer tribe who live along the banks of the Maici River, a tributary of the Amazon. Their language is the strangest in the world and has no known connection with any other living language. It is severely limited with no descriptive words but, even stranger, it has no counting words. The tribe feel no need to count or calculate. '*Hói*' is their closest word for 'one' but they also use it to mean a small amount. They don't even use their fingers for counting when they want to barter 4 baskets of Brazil nuts for 1 bottle of whisky.

2,500 years after Pythagoras, in our modern world of enormous computer-generated numbers, it is hard to believe that there are people who have no concept of 1.

2

2 is a favourite number. It is the number of love. When you sneeze, someone will say, 'One's a wish and two's a kiss …' Then there is 'Two's company, three's a crowd.' The Latin word for 2 is '*duo*', giving us 'dual carriageway', 'dual control' and many other words and phrases.

2 is the first even number and the first prime number.

The eccentric Greek mathematician Pythagoras and his equally nutty chums believed that the world could be understood through numbers, which they worshipped. They believed that human beings were reincarnated every 196 years. They also believed that even numbers were female and earthy, and odd numbers male and divine. 2 was the first female number.

The Russian Way to Multiply

Still on the subject of 2, if long multiplication isn't your forte but you are OK with multiplying and dividing by 2, try this: for multiplying, for example, 39 × 45, write the numbers in 2 columns and divide the first column by 2 repeatedly, ignoring remainders, until you end with 1. In the second column, multiply by 2 repeatedly until it matches the first column. In this column, score out any numbers that lie opposite an even number in the first column and add the remaining numbers in the second column.

39	45
19	90
9	180
4	~~360~~
2	~~720~~
1	1440

Total for 2nd column = 1755

The answer to 39 × 45 is 1755.

This works for any multiplication, but for a calculation such as 1276 × 244, the columns get very long! They used to do long multiplication this way in Russia but it's probably been forgotten now that everyone has a pocket calculator.

The Binary System

The binary system is not used in everyday calculation but almost everybody knows that binary numbers are used in computer science. However, as long as the computer is happy to accept data in our decimal number system and churn the data back out again in decimal number form, the average person is unconcerned about the binary numbers doing their secret stuff in the innards of the machine.

Like many mathematical discoveries, binary numbers were invented long before scientists found a practical use for them.

The binary system, where any number can be written using only the digits 1 and 0, was invented by the great German mathematician, Gottfried Leibniz.

Leibniz was a very religious man. He associated 1 with God and 0 with nothingness, and it pleased him that all numbers could be created out of unity and nothing.

The binary system is easy and enjoyable to learn and this is how it works.

Try to remember how you started to count. There were headings:

←	←	100s	10s	Units
		7	4	2

So $7 \times 100 = 700$, $4 \times 10 = 40$ and $2 \times 1 = 2$, giving 742.

The headings for binary numbers are:

$\leftarrow$ $\leftarrow$ 16s   8s   4s   2s   Units

So take the number 15, for example. This is $8 + 4 + 2 + 1$ so *one* lot of each of these goes under the appropriate headings giving the binary number 1111_2. The subscript '2' must always be attached to make it clear that this is not a decimal number.

Now take 22. This is $16 + 4 + 2$ and putting these under the appropriate headings, remembering to fill in the missing headings with 0, gives the binary number 10110_2.

The headings, of course, can be extended as far as you need by doubling the previous heading.

There is another way to convert a decimal number to a binary number. For example, take 55_{10}:

$55 \div 2 = 27 \text{ r } 1$
$27 \div 2 = 13 \text{ r } 1$
$13 \div 2 = 6 \text{ r } 1$
$6 \div 2 = 3 \text{ r } 0$
$3 \div 2 = 1 \text{ r } 1$
$1 \div 2 = 0 \text{ r } 1$

Write down the remainders from the bottom up: 110111

So $55_{10} = 110111_2$

Try another one. 48_{10}:

$48 \div 2 = 24 \text{ r } 0$

$$24 \div 2 = 12 \text{ r } 0$$
$$12 \div 2 = 6 \text{ r } 0$$
$$6 \div 2 = 3 \text{ r } 0$$
$$3 \div 2 = 1 \text{ r } 1$$
$$1 \div 2 = 0 \text{ r } 1$$

So, from the bottom up, $48_{10} = 110000_2$.

Changing back from binary to decimal is even easier. 111001_2 (using the headings) gives 1×64, 1×32, 1×16, 0×4, 0×2, $1 \times 1 = 113_{10}$.

Binary numbers get very long very quickly. The 3-digit 500_{10} has 9 digits in the binary system: 111110100_2.

A mathematical joke: There are 10 types of people in this world – those who understand binary numbers and those who don't.

The Match Secretary's Favourite Numbers

Sports match secretaries love the binary-number headings and in particular, the 16s and 32s, because these are the ideal entry numbers for knockout competitions. The first round of the tennis singles competition with 32 players is reduced immediately to 16, then to 8, then 4 and onwards to the grand final.

But a match secretary's job is not usually that easy. Suppose there are only 31 names on the entry sheet and no one else can be persuaded to enter. What does the match secretary do?

He subtracts 31 from the next binary heading, which is 32. So 1 person gets a bye into the second round. Meanwhile, the other 30 play the first round, ending with 15 winners who, with the person who got the bye, will make 16 going into the next round and the competition proceeds.

There is nothing more annoying than when, at the last minute, 7 would-be Andy Murrays add their names to the 32 on the list. What do you do with 39 entries?

This time, you go to 64s, the next binary heading. Subtracting 39 from 64 means that 25 players get a bye into the second round. Everyone has to wait for weeks in a particularly rainy summer for only $39 - 25 = 14$ players to play the first round, thereby allowing the $25 + 7 = 32$ to get on with the competition.

So the method is:

1. Subtract the number of entries from the next binary heading.
2. This number will get a bye.
3. The number of entries minus the number of byes will play the first round.
4. The winners of the first round plus the byes will proceed with the competition.

Odious and Evil Numbers

Surely not! What on earth have these numbers done to deserve titles like that?

The clue is in the first 2 letters of the words – 'od' and 'ev'. Ah, the penny is dropping – it's one of these mathematicians' jokes, isn't it? They do have a weird sense of humour (weird numbers are discussed in '70 and Other Weird Numbers' later in the book).

Odious numbers are binary numbers with an odd number of legs and evil numbers are, of course, binary numbers with an even number of legs.

Decimal number	Binary number
1	1
2	10
3	11
4	100
5	101
6	110
7	111
8	1000
9	1001
10	1010
11	1011
12	1100
13	1101
14	1110
15	1111
16	10000
17	10001
18	10010
19	10011
20	10100

So, for the numbers from 1 to 20, the odious numbers are 1, 2, 4, 7, 8, 11, 13, 14, 16 and 19, and the evil ones are 3, 5, 6, 9, 10, 12, 15, 17 and 18. The odious and the evils remain equal in number throughout the number system.

Square Numbers

This diagram above contains the square numbers 1, 4, 9 and 16. To increase the pattern, just add an extra back-to-front L-shape as shown. So the squares continue, 25, 36, 49, 64, 81, 100 ... and they go on forever. The squares of odd numbers are odd and the squares of even numbers are even. They can be written 1^2, 2^2, 3^2, 4^2, 5^2 and so on.

Pythagoras

Square numbers appear in maths more often than any other power. They become second nature to young teenagers who are studying the Theorem of Pythagoras at school.

$$5^2 = 3^2 + 4^2$$
$$13^2 = 5^2 + 12^2$$
$$17^2 = 8^2 + 15^2$$

Groups of Pythagorean triples like (3, 4, 5), (5, 12, 13) and (8, 15, 17) form right-angled triangles and the groups go on forever.

All these original triples can be expanded by multiplying so (3, 4, 5) can become (6, 8, 10) or (33, 44, 55) and a never-ending number of other multiples.

Far along the number line there is another original Pythagorean triple: (693, 1924, 2045). Get out your calculator and test it.

$2045^2 = 4,182,025$

and

$693^2 + 1924^2 = 480,249 + 3,701,776 = 4,182,025$

Hey presto!

Here is a right-angled triangle:

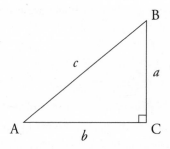

The formula for the area of a triangle is ½b × a. In our case, ½ of 1924 × 693 = 666,666 square units (and that looks like a very interesting number.)

How to Generate Pythagorean Triples
Using Simple Arithmetic

Choose any 2 fractions or a fraction and a whole number that multiply to 2. For example, $\frac{4}{3} \times \frac{3}{2}$ or $\frac{1}{5} \times 10$. Let's start with the first one.

> Add 2 to each fraction
> $2\frac{4}{3}$ and $2\frac{3}{2}$
> $= \frac{10}{3}$ and $\frac{7}{2}$
> Cross multiply to get 21 and 20
> These are the shorter sides.
> Now use $a^2 + b^2 = c^2$
> $21^2 + 20^2 = 441 + 400 = 841$
> $c = \sqrt{841} = 29$
> You have the Pythagorean triple (20, 21, 29)

Taking a shortcut with $\frac{1}{5}$ and 10, you have $2\frac{1}{5}$ and 12, which are $\frac{11}{5}$ and $\frac{12}{1}$. Cross multiply to get 11 and 60, use the $a^2 + b^2 = c^2$ formula to get 61 and the (11, 60, 61) triple. To generate (3, 4, 5), the triple everyone knows, start with 1 and 2. Add 2 and you have $\frac{3}{1} \times \frac{4}{1}$. Cross multiply to get 3 and 4, etc.

Pythagorean Triangles with the Same Perimeter

6 right-angled triangles each have a perimeter of 720. They are (180, 240, 300), (120, 288, 312), (144, 270, 306), (72, 320, 328), (45, 336, 339), (80, 315, 325). Although they have equal perimeters, they do not have equal areas.

Every Square Number is the Sum of Consecutive Odd Numbers

$1 = 1$
$4 = 1 + 3$
$9 = 1 + 3 + 5$
$16 = 1 + 3 + 5 + 7$
$25 = 1 + 3 + 5 + 7 + 9$

And on and on you go through the never-ending square numbers.

Another Pattern for Squares

$1 + 2 + 1 = 4 = 2^2$
$1 + 2 + 3 + 2 + 1 = 9 = 3^2$
$1 + 2 + 3 + 4 + 3 + 2 + 1 = 16 = 4^2$
$1 + 2 + 3 + 4 + 5 + 4 + 3 + 2 + 1 = 25 = 5^2$

And this big one:

$1 + 2 + 3 + 4 + 5 + 6 + 7 + 8 + 9 + 10 + 11 + 12 + 11 + 10$
$+ 9 + 8 + 7 + 6 + 5 + 4 + 3 + 2 + 1$
$= 144$
$= 12^2$

The beautiful pattern goes on forever.

$$X^2 - Y^2 = (X - Y)(X + Y)$$

Is this formula familiar from your schooldays? It is so useful for subtracting 2 squares when you've mislaid your calculator. For example:

$$67^2 - 63^2 = (67 - 63)(67 + 63) = 4 \times 130 = 520$$

and

$$989^2 - 11^2 = (989 - 11)(989 + 11) = 978 \times 1000 = 978000$$

Calculating Squares in Your Head

Doing sums in your head without the use of paper and pencil or pocket calculator will mightily impress your friends. Here is an easy way to calculate 15^2, 25^2, 35^2 and many more.

Take the digit or digits before the 5, add on 1 and multiply. Then attach 25. So for 45^2, take the 4, add on 1 and multiply 4×5, and attach 25, getting 2025. And for 95^2, take the 9, add on 1 and multiply 9×10, and attach 25, getting 9025.

Bask in your friends' admiration as you calculate 9995^2 (999×1000 and attach 25) in 2 seconds flat. Adopt a pose of modest intelligence. Mm – that's 99,900,025, isn't it? It's even better if you say it in words. Now you know $225 = 15^2$ and $625 = 25^2$ but you can add 225,625 to your repertoire: $225,625 = 475^2$.

The Cube of a Square is Always a Square

$(2^2)^3 = 4^3 = 64 = 8^2$
$(3^2)^3 = 9^3 = 729 = 27^2$
$(11^2)^3 = 121^3 = 1,771,561 = 1331^2$
$(25^2)^3 = 625^3 = 244,140,625 = 15,625^2$

This always works because of a simple little bit of algebra: $(n^2)^3$ = n^6 and $n^6 = n^3 \times n^3$.

Take Any 2 from 4

p q, r and s are 4 numbers: 10,430, 3970, 2114 and 386.

$p + q = 10,430 + 3970 = 14,400 = 120^2$
$p + r = 10,430 + 2114 = 12,544 = 112^2$
$p + s = 10,430 + 386 = 10,816 = 104^2$
$q + r = 3970 + 2114 = 6084 = 78^2$
$q + s = 3970 + 386 = 4356 = 66^2$
$r + s = 2114 + 386 = 2500 = 50^2$

And all 4 numbers add to $16,900 = 130^2$.

Twin Primes

Twin primes are pairs of consecutive odd numbers that are both prime. In the first 100 numbers, there are 8 pairs: (3, 5), (5, 7), (11, 13), (17, 19), (29, 31), (41, 43), (59, 61), (71, 73), and they go on and on to the largest pair found so far with over 200,000 digits. Every pair has a multiple of 6 between them except (3 and 5), which, because they started it all, don't always have to obey the rules.

But do these twin pairs go on forever like the single primes? Well, that is not known for certain. Most mathematicians think they probably do although no proof has yet been found.

In 1737 Leonhard Euler proved that the sum of the reciprocals of the prime numbers ($\frac{1}{2}$ + $\frac{1}{3}$ + $\frac{1}{5}$ + $\frac{1}{7}$ + $\frac{1}{11}$ + $\frac{1}{13}$ + $\frac{1}{17}$ + $\frac{1}{19}$, etc) must diverge. This means that as you extend this series the sum keeps rising without limit.

Viggo Brun, the twentieth-century Norwegian mathematician, investigated the sum of the reciprocals of the twin pairs: ($\frac{1}{3}$ + $\frac{1}{5}$) + ($\frac{1}{5}$ + $\frac{1}{7}$) + ($\frac{1}{11}$ + $\frac{1}{13}$) + ($\frac{1}{17}$ + $\frac{1}{19}$) + and on and on to the mega twin prime reciprocals.

However, Brun proved that this sum converges, that is, there is a point above which it will never rise. This point is very difficult to calculate but is thought to be around 1.90216058 … but maybe not exactly!

So this method doesn't prove that the twin prime pairs are infinite, and while most mathematicians believe that they probably are, it remains unproven.

2 Interesting for Words

$(3^2 + 4^2 + 5^2 + 6^2 + 7^2 + 8^2 + 9^2) \div (1^2 + 2^2 + 3^2 + 4^2 + 5^2 + 6^2 + 7^2) = 280 \div 140 = 2$

Growing Primes

2 can grow 7 times and still be prime. The following are prime and have been checked!

2
29
293
2939
29,399
293,999
2,939,999
29,399,999

But all good things come to an end: 293,999,999 is divisible by 29.

Indices

In maths, 'indices' is the plural of 'index'. In the calculation 3^4 = $3 \times 3 \times 3 \times 3$ = 81, the little power number 4 is the index figure.

In the Pythagoras calculations a few pages back, you had to add $20^2 + 21^2$. There is no quick way of doing this. You cannot, for example, add 20 and 21 and square that. Apart from the very handy formula for finding the difference of 2 squares, $X^2 - Y^2 = (X - Y)(X + Y)$, there is no quick way of adding or subtracting powers of numbers.

So $3^2 + 3^3 + 3^4 + 3^5$ has to be tackled head on.

However, multiplying and dividing indices is different. $a^m \times a^n = a^{m + n}$ means that if you have to calculate $3^2 \times 3^3 \times 3^4$, you might find it quicker to calculate 3^9. It is certainly easier for dividing. $a^m \div a^n = a^{m - n}$ means that if you have to calculate, say, $7^{11} \div 7^9$, all you need to do is subtract the index numbers and calculate 7^2 = 49, thus saving half an hour of your precious time. $4^4 \times 4^6 \div 4^8$ should take you 3 seconds. (Answer 16.)

$(a^m)^n = a^{mn}$ means that when brackets are involved, you multiply. So $(2^3)^2 = 2^6 = 64$.

You may come across a negative index like 3^{-2}. Don't worry, that simply means 'one over'. So 3^{-2} = 1 over 3^2 = ⅑.

You can also do a bit of controlled mixing up. There is no quick way to calculate $2^3 \times 4^2$. You have to work them out separately and get 8×16 = 128. But you can do $2^3 \times 4^3$ (because they have the same index figure) and get 8^3 = 512 instead of, separately, 8 × 64 = 512. Likewise, 2^6 × 5^6 = 10^6 = 1,000,000. Doing this the long way takes ages. 2^6 = 64 and 5^6 = 15,625 and 64 × 15,625 = 1,000,000.

Fractional Indices

$a^{1/2}$ looks scary but it means exactly the same as $\sqrt{a}$, the square root of a. Likewise, $a^{1/3}$ means the cube root of a number. So $25^{1/2}$ = 5 and $1000^{1/3}$ = 10.

$a^{2/3}$ means the number needs to be squared first, then cube rooted or, if you wish, cube rooted first and then squared. In fact, it is usually quicker to do the cube root first. For example, calculate $27^{2/3}$. The cube root of 27 is 3 and squaring gives 9. The other way means squaring 27 and then finding the cube root of 729. The answer is 9 again, of course, but it takes much longer.

Standard Form

Mathematicians and scientists often have to work with very large or very small numbers: for example, the approximate speed of light, 300,000,000 metres per second, or the mass of a tiny particle, 0.0000014 grams.

Typing the correct number of zeros would make their secretaries quite cross-eyed and could result in serious errors, so mathematicians invented standard form, a method of expressing these awkward numbers in a more concise form, which avoids writing strings of zeros.

Each number is written in the form a $\times$ 10^n where a is a number between 1 and 10 and n, the index figure, is a positive or negative whole number. A few examples:

1. 18,000,000: Take the '18' bit, and insert a decimal point between the 1 and 8 to make 1.8 (a number between 1 and 10), then count the leaps from the point to the end. In this case there are 7 leaps, so 18,000,000 = 1.8×10^7

2. 50,000: This time the number between 1 and 10 is 5.0 and there are 4 leaps to the end, so 50,000 = 5.0×10^4

3. 1 googol (that's the name of the huge number with 100 zeros) = 1.0×10^{100}

For very small numbers, you do much the same but leap to the left until you reach the original decimal point.

4. $0.00000034 = 3.4 \times 10^{-7}$
5. $0.006 = 6.0 \times 10^{-3}$

Advice: leave standard form to the mathematicians. It is rather pretentious to say that your new sports car cost £9.50×10^4.

A Twentieth-century Conjecture

A fun, modern conjecture was proposed by Lothar Collatz, the renowned German mathematician, who died in 1990.

He invited you to choose any whole number. Divide it by 2 if it is even or multiply it by 3 and add 1 if it is odd and keep on going. Eventually, you will end up with 1.

For starters, let's choose 7. So we have:

7→22→11→34→17→52→26→13→40→20→10→5→
16→8→4→2→1

And another one:

23→70→35→106→53→160→80→40→20→10→5→
16→8→4→2→1

And let's get really ambitious:

1000→500→250→125→376→188→94→47→142→71→
214→107→322→161→484→242→121→364→182→91→
274→137→412→206→103→310→155→466→233→
700→350→175→526→263→790→395→1186→593→
1780→890→445→1336→668→334→167→502→251→

754→377→1132→566→283→850→425→1276→638→
319→358→179→538→269→808→404→202→101→
304→152→76→38→19→58→29→88→44→22→11→
34→17→52→26→13→40→20→10→5→16→8→4→2→1

Phew! Never again! You need to have a lie-down after that! However, if you want a short one to demonstrate the conjecture, just select any of the numbers near the end of the above calculation. No one has yet found a starting number that didn't work – eventually! We'll take their word for it.

3

3 is a much-used number in the English language: 3 blind mice, the 3 wise men, the 3 Rs, and a drunken sailor is said to be '3 sheets to the wind'. And did you know that an octopus has 3 hearts? Also, before we forget, 3 cheers for the inventor of the pocket calculator.

3 is the first odd prime number and, to the Pythagoreans, 3 was male and divine.

3 is associated with the triangle and the branch of mathematics called trigonometry that deals with the sides and angles of a triangle.

Draw any triangle and construct the medians (the lines drawn from the vertices of the triangle to the midpoint of the opposite sides).

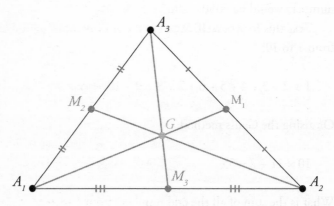

3s again! The point G is a point of trisection on the medians. This means that, for all 3 medians, AG is ⅔ of AM and GM is ⅓ of AM.

Karl Friedrich Gauss

Karl Friedrich Gauss (1777–1855) was one of the world's greatest mathematicians.

When he was a young boy, he and his classmates were told to add up all the numbers from 1 to 100. Their teacher, Herr Buttner, was tired and he hoped this task would keep the boys quietly occupied for the rest of the afternoon. But after a few seconds, the little genius raised his hand and said, 'The answer, sir, is 5050.' He had multiplied 100 by 101 and divided by 2. He had, in fact, reinvented the triangular numbers first discovered by the Ancient Greeks.

If young Karl had been asked to add up all the odd numbers between 1 and 100, that would have been equally easy for him: $(100 \div 2)^2 = 50^2 = 2500$. So, of course, the sum of all the even numbers would be $5050 - 2500 = 2550$.

Test this for yourself. What is the sum of all the numbers from 1 to 10?

$$1 + 2 + 3 + 4 + 5 + 6 + 7 + 8 + 9 + 10 = 55$$

Or, using the Gauss method:

$$10 \times 11 \div 2 = 55$$

What is the sum of all the odd numbers from 1 to 10?

$$1 + 3 + 5 + 7 + 9 = 25 \text{ or } (10 \div 2)^2 = 25$$

The sum of the even numbers is:

$$2 + 4 + 6 + 8 + 10 = 30 \text{ or } 55 - 25 = 30$$

The Triangular Numbers

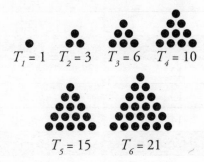

The triangular numbers go on forever because, in the images above, you just need to add a new row of spots. The fifth number is 15. You can count the spots, $1 + 2 + 3 + 4 + 5 = 15$, or you can do the calculation the Gauss way, $5 \times 6 \div 2 = 15$. The 500th triangular number is $500 \times 501 \div 2 = 125{,}250$. You can easily invent the formula:

$$n(n + 1) \div 2$$

Here is the beginning of the list of triangular numbers:

1, 3, 6, 10, 15, 21, 28, 36, 45, 55, 66, 78, 91, 105, 120, 136, 153, 171, 190, 210 …

Gauss also discovered that every whole number is the sum of no more than 3 triangular numbers. (1 also gets to be included as it claims to be the first triangular number – although anyone can see that it isn't a triangle.)

Whole number	Triangular numbers
1	1
2	1, 1
3	3
12	6, 6
13	3, 10
14	1, 3, 10
55	10, 45
56	1, 55
57	21, 36
999	3, 6, 990
1,000	6, 91, 903
1,001	10, 171, 820

Like every mathematician, Gauss was also interested in prime numbers and after a day of heavy mathematics, he would relax by spending a pleasant half hour or so counting a chiliad of primes. (A chiliad is a group of 1,000 things pronounced with a 'k' as in

Kit Kat, the chocolate bar.) He estimated that in all these leisurely moments he had counted up to 3 million prime numbers and he was able to conjecture about their distribution and density.

Gauss made significant contributions to most fields of mathematics and is one of the greatest in the history of the subject.

His brain, along with that of Dirichlet, is preserved in the physiology department of the University of Göttingen. Gauss's brain weighs 1492 grams, several grams heavier than that of Dirichlet.

Divisibility by 3

After the work of a genius, here is something easy. How can you tell at a glance if any number divides by 3? Simple – just add up the digits and if this is a number on the 3 times table, then the original number divides by 3. For example, 141, 882, 11,778 and 111,111 all divide exactly by 3, because the digit totals are 6, 18, 24 and 6.

Square Numbers and Triangular Numbers

Every square number from 2^2 onwards is the sum of 2 consecutive triangular numbers. (Reminder, the triangular numbers are 1, 3, 6, 10, 15, 21, 28, 36, 45, 55, 66 and so on.)

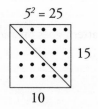

$5^2 = 25$

15

10

The diagram shows that the 5 by 5 square is composed of T_4 (10) and T_5 (15). The equations below can be demonstrated by similar squares sliced into triangles.

$2^2 = 1 + 3$
$3^2 = 3 + 6$
$4^2 = 6 + 10$
$5^2 = 10 + 15$
$6^2 = 15 + 21$
$7^2 = 21 + 28$

And so on. All along the triangular number list beginning 1, 3, 6, 10, 15, 21, 28, 36, 45, 55, 66 and 78, every 2 consecutive numbers add to a square number and that goes on forever.

8 Times a Triangular Number Plus 1 is Always a Square Number

$8 \times 1 + 1 = 3^2$
$8 \times 3 + 1 = 5^2$
$8 \times 6 + 1 = 7^2$
$8 \times 10 + 1 = 9^2$
$8 \times 15 + 1 = 11^2$
$8 \times 21 + 1 = 13^2$

And so the beautiful pattern goes on and on forever.

3 Dice

Get a friend to roll 3 dice (or the same one 3 times) and ask them to hide the results. Ask him to double the number of spots on the first dice, add 5, then multiply by 5. Add on the number of spots on the second dice and then multiply by 10. Finally, add on the 3rd dice. When he tells you the total, subtract 250 and you know what the 3 dice were.

For example if the 3 results were 4, 1 and 6, the calculation would be:

$$[(4 \times 2 + 5) \times 5 + 1] \times 10 + 6 = 666$$

Now subtract 250 and you get 416.

Actually this works for any number from 1 to 9.

An Extraordinary 3 × 3 Magic Square

A magic square is a square array of numbers where the rows, columns and diagonals add up to the same number. This one is no ordinary magic square. It is la crème de la crème of all 3 × 3 magic squares. It was invented by the renowned American mathematician, Harry L. Nelson, in response to a competition sponsored by fellow mathematician Martin Gardner in 1988. It may only be a mere 3 × 3 square but it contains nine 10-digit prime numbers, all of which are close neighbours far along the number line.

1480028201	1480028129	1480028183
1480028153	1480028171	1480028189
1480028159	1480028213	1480028141

Nelson's prize was only $100.

No other 3 × 3 magic square can compare with Nelson's square. Certainly not this boring one below.

A Boring Little Magic Square that Becomes Interesting

5	22	18
28	15	2
12	8	25

However, draw another square with the numbers written in words.

five	twenty-two	eighteen
twenty-eight	fifteen	two
twelve	eight	twenty-five

Now count the letters in each cell and make another new square.

4	9	8
11	7	3
6	5	10

And that is nice.

Triple Primes

Triple primes have much the same definition as twin primes. They are 3 consecutive odd numbers that are prime. They are (3, 5, 7) … and … well, that's it! There are no more – ever! You see, in every 3 consecutive odd numbers, one of them is divisible by 3. Think about it. In (17, 19, 21) and (61, 63, 65) and (105, 107, 109), 21, 63 and 105 are multiples of 3. In other words, in every trio of odd numbers there is always one of them that is not prime. In the first and only trio of primes, 3 is acting in its role of first odd prime number.

Prime Triplets

Apart from the one and only group of triple primes above, the closest grouping of 3 primes is of the form $(p, p + 2, p + 6)$ or $(p, p + 4, p + 6)$, such as, for example, (11, 13, 17), (67, 71, 73), (97, 101, 103) or (881, 883, 887).

They are called prime triplets and, in every case, 2 of the triplets are actually twins and the other triplet is a bit of an outsider, being 6 more than the youngest twin or 6 less than the oldest twin.

There are also cousin primes of the form $(p, p + 4)$ like (37, 41) and (79, 83) and prime quadruplets of the form $(p, p + 2, p + 6, p + 8)$. Each of these quadruplets contains 2 sets of twins such as (101, 103, 107, 109).

If a quadruplet like the one above has either $p - 4$ or $p + 12$ prime, then there is a prime quintuplet such as (97, 101, 103, 107, 109) or (101, 103, 107, 109, 113).

If the quadruplet has both $p - 4$ *and* $p + 12$ prime, then you have a prime sextuplet. The only 2 simple ones are

3

(7, 11, 13, 17, 19, 23) and (97, 101, 103, 107, 109, 113). The next set are unfamiliar 5-digit primes a long way off along the number line.

Cubes and Cube Roots

Most people know their squares by heart, at least up to 10. Not so many know their cubes, but they are not so very difficult to remember.

Number n	1	2	3	4	5	6	7	8	9	10
Cube n^3	1	8	27	64	125	216	343	512	729	1000

There are 21 cubes under 10,000. The last one is $21^3 = 9261$.
Note:

1. Every cube from 1 to 10 ends with a different digit.
2. 4^3, 5^3, 6^3, 9^3 and 10^3 all end with a matching digit.

The symbol for a cube root is $\sqrt[3]{}$ and so $\sqrt[3]{729} = 9$.

Cube roots are not found on simple pocket calculators but it's surprising how well behaved cube roots can be. Keep in mind:

$10^3 = 1000$
$20^3 = 8000$
$30^3 = 27,000$
$40^3 = 64,000$
$50^3 = 125,000$
$60^3 = 216,000$

$70^3 = 343,000$
$80^3 = 512,000$
$90^3 = 729,000$

$\sqrt[3]{4096}$ is between 1000 and 8000 in the above list, and since 4096 ends in 6 (see the table above) the cube root is likely to be 16. Check $16 \times 16 \times 16 = 4096$.

Now, suppose you want to calculate $\sqrt[3]{250,047}$. Examine the '250' bit. It comes between 216,000 and 343,000 above. Choose the lower one so your cube root answer is going to be '60 something'. Now look at the last digit in your number. It is 7 and in the table the only '7' is part of 27, the cube of 3. Check on your calculator, $63 \times 63 \times 63 = 250,047$. So $\sqrt[3]{250,047} = 63$.

Let's try it once again. What is $\sqrt[3]{658,503}$? Look at the '658' bit. It comes between 512,000 and 729,000 above. Choose the lower number so your cube root answer will be '80 something'. Look at your number again. The last digit is 3 and in the table above, the only 3 is part of 343, the cube of 7. So $\sqrt[3]{658,503} = 87$.

Cube Roots You Will Like

To calculate the cube root of 125, ask 'What times what times what gives 125?' The answer is 5. But there are a few special cube roots where you can abandon the 'what times what times what' business.

They are 5 numbers whose cube roots are obtained by merely adding the digits:

$\sqrt[3]{512} = 8$
$\sqrt[3]{4913} = 17$

$$\sqrt[3]{5832} = 18$$
$$\sqrt[3]{17,576} = 26$$
$$\sqrt[3]{19,683} = 27$$

Remember, these are the only ones! For all the others, you have to do it the hard way.

Cute Cubes

$$3^3 + 4^3 + 5^3 = 6^3$$

Mersenne Numbers

You might be forgiven for thinking that mathematicians are obsessed with prime numbers.

Marin Mersenne, the seventeenth-century monk, philosopher and mathematician, searched for a formula to generate prime numbers. For all of five seconds, he might have considered the formula $2^n - 1$. Yes, when n = 2, $2^2 - 1 = 3$ and 3 is a prime number. Likewise, $2^3 - 1 = 7$ is a prime number. But the formula fails for $2^4 - 1$ because 15 is not a prime number.

Then he thought, 'Aha, but what if I limit n to prime numbers only? If p is prime, will $2^p - 1$ also be prime?' Well, the answer is, sometimes it is, but often it isn't.

$$2^5 - 1 = 31$$
$$2^7 - 1 = 127$$
$$2^{11} - 1 = 2047$$

Alas $2047 = 23 \times 89$ so $2^{11} - 1$ is not prime.

$2^{13} - 1 = 8191$, $2^{17} - 1 = 131071$ and $2^{19} - 1 = 524287$ are all prime – don't worry, someone has already checked them. But it fails again at $2^{23} - 1$ and $2^{29} - 1$, picks up at $2^{31} - 1$ but immediately fails again. Then there is a huge gap before it restarts at $2^{61} - 1$. This enormous prime number has 19 digits but the biggest prime found so far, the 48th Mersenne prime, has nearly 17 million digits! Printing out this whopping mega-number would take more than 4,500 pages of A4 paper, with 75 digits per line and 50 lines per page.

These numbers are unmanageable by everyone and everything except the most powerful supercomputers. But mathematicians are a tenacious lot so the battle to increase and verify the Mersenne primes continues to the present day.

Pierre de Fermat was a friend of Mersenne and, of course, he was also interested in prime numbers. He conjectured that the number $2^n + 1$ was always prime *if n was a power of 2*. It worked for n = 2, 4, 8 and 16, giving prime numbers 5, 17, 257 and 65,537, but 100 years later, Euler showed that $2^{32} + 1 = 4,294,967,297$ was not prime because it was divisible by 641.

A Spectacular Number Trick

A little light relief is required after all these mega-primes.

You invite your victim to write down any 3-digit number. Dissuade any smart-pants from writing 999. Suppose they choose 357. You decide to choose 357 also. They are then invited to choose another number, say 428, but this time, with much indecision and faffing about, you eventually choose the 999 complementary number 571 (they chose 4, you chose 9 − 4, which is 5, they chose 2, you chose 9 − 2, which is 7, they chose 8, you chose 9 − 8, which is 1), hoping that they will not notice your well-thought-out choice. Now hand them a calculator and instruct them to multiply their chosen numbers and your chosen numbers. Then they have to add both products. So their calculation is:

$(357 \times 428) + (357 \times 571)$
$= 152{,}796 + 203{,}847$
$= 356{,}643$

Now, of course, you can do this in your head. Subtract 1 from the first number, getting 356, and attach its 999 complementary number, 643, getting 356,643.

Fermat's Last Theorem

No 3 positive numbers, a, b and c, exist that satisfy the equation $a^n + b^n = c^n$ for any whole number value of n greater than 2.

This statement was famously written in 1637 by Pierre de Fermat in the margin of the book he was reading at the time, *Arithmetica*, by the third-century BC mathematician Diophantus.

Fermat added that he did have the proof but it was too large to fit in the margin. His proof was never found and, for the next 358 years, it was a point of honour for almost every mathematician to prove the statement but no one succeeded.

Fermat's conjecture was not of tremendous mathematical importance but it had caught the imagination of others who were not even mathematicians. Fermat's Last Theorem has appeared in fiction on many occasions including *Doctor Who*, *Star Trek*, *The Simpsons* and Stieg Larsson's novel *The Girl Who Played with Fire*, the second in his Millennium series.

But, at last, in 1995 the problem was finally solved by Sir Andrew Wiles, now Royal Society Research Professor at Oxford University, to worldwide acclaim.

4

4 is a commonly used number in the English language. The remotest parts of the world are called the four corners of the earth and motorists may even get there driving 4 × 4s. Four-letter words are rude and usually replaced in writing with ****.

It's a lucky day if you find a four-leafed clover. However, a gift of a four-leafed clover to a citizen of Beijing would cause the same reaction as the 'gift' of a horse's head from the Mafia. The number is unlucky throughout China because the Mandarin word for four sounds very like the Mandarin word for 'death'. Lifts in high buildings do not have a floor numbered four. Instead, it is called the 'upper third floor'. Even so, it's mostly unwitting foreign tourists who are packed into the rooms on that particular floor.

Four as a number is not that interesting but it is the source of a few number tricks. Here is one of them.

A Trick that Always Ends in Four

Think of a number and write it down in words. What about thirty-seven? That has eleven letters, so eleven will be the next number:

Thirty-seven (11 letters)
Eleven (6)

Six (3)
Three (5)
Five (4)
Four (4)

And 'four' for evermore.

Try it for yourself. No one has yet found a number that doesn't work. (The *second* last number may be five or nine but the last one will always be four.)

4 is Upstairs

$3^4 \times 425 = 34425$

4 Leads to an Infinite Number of Squares

$4^2 = 16$
$34^2 = 1156$
$334^2 = 111556$
$3334^2 = 11115556$
$33334^2 = 1111155556$

And the pattern goes on forever.

Divisibility by 4

Any number is divisible by 4 if the last 2 digits are on the 4 times table. For instance, 71<u>36</u> is divisible by 4 but 2,848,4<u>18</u> is not.

Tetrahedral Numbers

A regular tetrahedron, or to use its other name, triangular pyramid, has 4 identical equilateral faces.

In days gone by, wars were fought with cannons, usually mounted on wheels. From them, cannonballs were fired. These extremely heavy balls, made of stone or iron, were stacked ready to be fired in a tetrahedron shape.

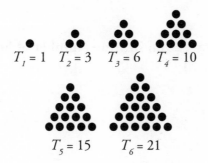

For example, the bottom layer of the triangle might have 21 balls. The next layer would have 15 balls fitting into the spaces, then 10 on top, then 6 and 3 with 1 at the top. The total number of balls is 56, and 56 is the 6th tetrahedral number.

So the tetrahedral numbers are made by adding the triangular numbers, starting, of course, with 1.

$$1$$
$$1 + 3$$
$$1 + 3 + 6$$
$$1 + 3 + 6 + 10$$
$$1 + 3 + 6 + 10 + 15$$
$$1 + 3 + 6 + 10 + 15 + 21$$
$$1 + 3 + 6 + 10 + 15 + 21 + 28$$
$$1 + 3 + 6 + 10 + 15 + 21 + 28 + 36$$

And on and on forever.

The tetrahedral numbers from the above triangle are: 1, 4, 10, 20, 35, 56, 84 and 120. A list of these numbers appears in the 4th diagonal row of Pascal's Triangle (see 'Pascal's Triangle' chapter later in the book).

The formula to generate any tetrahedral number is $n(n + 1)(n + 2) \div 6$. The 6th tetrahedral number is $6(6 + 1)(6 + 2) \div 6 = 6 \times 7 \times 8 \div 6 = 56$ (as shown above). The 20th tetrahedral number is $20(20 + 1)(20 + 2) \div 6 = 20 \times 21 \times 22 \div 6 = 1540$.

5

5 is another number much loved by the weird Pythagoreans. To them, 5 was the symbol of marriage because 5 = 3 + 2 and 3 is the first male number and 2 is the first female number.

Apart from that, human beings have five senses, Jesus fed the multitudes from five loaves (and two little fishes), and the vast five-sided building in Washington, DC, where 23,000 US government officials work, is called the Pentagon.

This figure is called a pentagram. Once the five 72-degree points are marked out on the circumference of a circle, the pentagram can be drawn in one continuous path of 5 straight lines. It is an important symbol in many religions and cults. The Pythagoreans, of course, declared it to be mathematical perfection, especially as there is another perfect little pentagon inside it and a larger pentagon if you join the points.

Occultists believe that the pentagram keeps away devils and witches, and that the top point represents heaven while the other 4 points represent earth, air, fire and water.

Multiplying by 5 the Easy Way

Multiplying by 5 isn't difficult but, without paper and pencil, can you do 6138842 × 5?

It's 30694210. Without multiplying and all that 'carrying' stuff, all you need to do is divide by 2 and attach 0.

Confirm these:

1.   446 × 5 = 2230
2.   4098 × 5 = 20490
3.   81766 × 5 = 408830

It is just as easy if your number is odd. Divide by 2 as before but this time attach 5. For example, 47243 × 5 = 236215.

Confirm these:

1.   445 × 5 = 2225
2.   12099 × 5 = 60495
3.   987654321 × 5 = 4938271605

The Pentagonal Numbers

You've had square numbers, prime numbers and triangular numbers. So what are pentagonal numbers? Well, here they are – 5-sided figures packed inside 5-sided figures. To get the numbers, you count the dots. And, of course, there is 1 as usual claiming its place in the series.

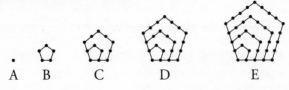

A   B   C   D   E

Everyone can see that 1 is just a miserable little dot and looks nothing like a pentagon but it insists on being there, so, counting the spots, the pentagonal numbers start 1, 5, 12, 22, 35, 51 … and they go on forever.

If the black spots are hurting your eyes, there's an easy formula to find any pentagonal number you wish:

½ n (3n – 1)

Let's confirm the number in image D. This time n = 4, so 2(12 – 1) = 22. Yes, there are 22 spots in D.

What about that huge pentagon, chock-a-block with other pentagons, way along the line with a side of 20 units. How many spots will that be?

The 20th pentagonal number is 20 × (3 × 20 – 1) ÷ 2 = 590.

Pentagonal Numbers and Triangular Numbers

Let's write out some of the pentagonal numbers again:

1, 5, 12, 22, 35, 51, 70, 92, 117, 145, 176, 210 …

and multiply them all by 3, giving:

3, 15, 36, 66, 105, 153, 210, 276, 351, 435, 528, 630 …

Now look again at the list of triangular numbers:

1, <u>3</u>, 6, 10, <u>15</u>, 21, 28, <u>36</u>, 45, 55, <u>66,</u> 78, 91, <u>105,</u> 120, 136, <u>153,</u> 171, 190, <u>210</u> …

Starting at 3, look at every third one. So every pentagonal number is one-third of a triangular number. Neat, isn't it?

The Fibonacci Sequence

In 1202, the most famous medieval mathematician, Fibonacci, was given a problem to solve. If 2 newly born rabbits, 1 male and 1 female, are allowed to breed in ideal conditions, how many rabbits will there be at the end of 12 months? (We have to assume that none of the rabbits die, they never escape from the field and always produce 2 babies, 1 boy bunny and 1 girl bunny, and that sibling mating is acceptable.)

So this is what happened during the procreation process:

After 1 month, there is 1 pair, ① and ❶.
After 2 months, there is still only 1 pair, ① and ❶, but they mate.
At 3 months, babies are born so there are now 2 pairs, ① and ❶, and ② and ❷.
At 4 months, babies ② and ❷ are too young but ① and ❶ produce ③ and ❸.
At 5 months, ① and ❶ have ④ and ❹ and ② and ❷ have ⑤ and ❺ (③ and ❸ are too young at this stage)

Now, if this is blowing your mind, this is what we have so far:

End of month	1	2	3	4	5
No of pairs	1	1	2	3	5

Fibonacci checked further, of course, but it didn't take him long to notice the pattern – each number was the sum of the previous 2 numbers. So the sequence continues: 1, 1, 2, 3, 5, 8, 13, 21, 34, 55, 89, 144, 233, 377, 610 ... and so on for ever. The twefth number along the line is 144 so after one year there will be 144 rabbits in the field.

Points to notice:

1. The sum of all the numbers along the line starting at the beginning will always equal 1 less than the number which is the next but one term. For example,

   $$1 + 1 + 2 + 3 + 5 + 8 + 13 + 21 = 55 - 1$$

   and

   $$1 + 1 + 2 + 3 + 5 + 8 + 13 + 21 + 34 + 55 + 89 + 144 + 233 = 610 - 1$$

2. Twice any term added to the term before equals the next but one term. For example:

   $$2 \times 89 + 55 = 233$$

   and

   $$2 \times 233 + 144 = 610$$

Fibonacci knew his maths but not his rabbits. The gestation time is indeed 1 month but the breeding season is usually only

9 months. Yes, really! They take 3 months off from their busy schedule! However, the average size of the litter is 5 or 6 but 8 or more is not uncommon. Now, how about inventing a new sequence over only 9 months but assuming the litter size is 6, 3 bucks and 3 does?

The Pythagoras Connection

About seventeen centuries separate Pythagoras and Fibonacci but a magic little bit of maths brings them together. Let's write out the sequence again:

1, 1, 2, 3, 5, 8, 13, 21, 34, 55, 89, 144, 233, 377, 610, 987, 1597, 2584 …

Choose any 4-number block, say 8, 13, 21, 34.

Multiply the outer ones, 8 × 34 = 272, and multiply the inner ones and double your answer, 13 × 21 × 2 = 546. Let 272 and 546 be the shorter sides of a right-angled triangle. Now calculate the length of the hypotenuse by applying the Theorem of Pythagoras.

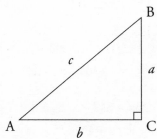

$$c^2 = a^2 + b^2$$
$$= 272^2 + 546^2$$
$$= 73984 + 298116$$
$$= 372100$$
$$c = \sqrt{372100}$$
$$= 610$$

(And you'll notice 610 further along the sequence.)

But the magic isn't over yet. The area of a triangle is ½ base × height:

$$= ½ \times 272 \times 546$$
$$= 74256 \text{ square units}$$

Now multiply the 4 Fibonacci numbers you first chose, 8 × 13 × 21 × 34, and look what you get. Pure magic, isn't it?

Divisors and Primes with Fibonacci

This is perhaps less spectacular, but interesting all the same.

Write out a small part of the Fibonacci sequence once more for your convenience:

1, 1, 2, 3, 5, 8, 13, 21, 34, 55, 89, 144, 233, 277, 610, 987, 1597, 2584, 4181, 6765, 10946, 17711, 28657 …

Write down any 2 numbers where one is divisible by the other, say 15 and 5. Now count along to the 15th Fibonacci number (we will call it F_{15}) and find also F_5. They are 610 and 5, and 610 ÷ 5 = 122. Try it again with 16 and 8. F_{16} = 987 and F_8 = 21 and 987 ÷ 21 = 47.

Now find prime number 233 in the sequence. It is F_{13} and 13 is prime. Look for 13 and it is F_7 and 7 is prime. Likewise, 1597 is prime and the subscript number in F_{17} is prime. In other words, if the Fibonacci number is prime so is the matching subscript number. The only exception is at the beginning where 3 is F_4 and 4 is not prime.

But it doesn't always work the other way round. The subscript number in F_{19} is prime but the 19th number along the Fibonacci list, 4181, (37 × 113) is not prime.

So all Fibonacci prime numbers have Fibonacci prime number subscripts but not all Fibonacci prime number subscripts yield a Fibonacci prime number. Got that?

Highest Common Factor (HCF) with Fibonacci

You may have forgotten what HCF is. The factors of 20 are 1, 2, 4, 5 and 10, and the factors of 12 are 1, 2, 3, 4 and 6, so the highest factor common to them both, the HCF, is 4.

Now find F_{20}, F_{12} and F_4 on the list above. They are 6765, 144 and 3. The factors of 6765 are 3, 5, 11, 41 and 144 is $2^4 \times 3^2$ so the HCF of 6765 and 144 is only 3. Try it again with your own choice of numbers. It always works.

Miles and Kilometres with Fibonacci

The Fibonacci sequence is a handy miles/kilometre converter. The approximate answer is always the next number in the sequence:

8 miles ↔ 13 km
21 miles ↔ 34 km
89 miles ↔ 144 km

A Truly Magical Reciprocal

Here is the reciprocal of 998,999:

$$1 \div 998{,}999 = 0.000001001002003005008013021034055089144233377610 \ldots$$

Compare it with the list of Fibonacci numbers:

0, 1, 1, 2, 3, 5, 8, 13, 21, 34, 55, 89, 144, 233, 377, 610 …

With zeros taken out and some spaces added, there it is – the Fibonacci sequence in all its glory.

The Flowers

The Fibonacci sequence has many mathematical applications but there is one very strange phenomenon. Many flowers have a Fibonacci number of petals. The clover has 3 leaves (unless it is a lucky 4-leaved clover), a lily has 3 petals, buttercups have 5, pansies 5, petunias 5, delphiniums 8, marigolds 13 and asters 21. The black-eyed Susan plant has 21 and different varieties of the daisy family have 21, 34 or 55.

A recent *Sunday Times* article reminds us that Alan Turing, breaker of the Nazis' Enigma code, conjectured in a scientific paper in 1952 that patterns on sunflower seedheads were linked to the Fibonacci number sequence. However, his research was left unfinished as he died two years later. To mark the centenary of his birth, the Sunflower Project, sponsored by the Museum of Science and Industry in Manchester, has restarted the research and over 10,000 people from several countries have donated sunflower specimens. It has been discovered that 82 per cent of the seeds examined so far had their seedheads arranged in clockwise and anticlockwise spiral patterns that followed the Fibonacci sequence.

Sceptics point to other flowers with non-Fibonacci numbers of petals but mathematicians believe the magic.

More Magic with Fibonacci

Here is a magic square that would insult a six-year-old.

2	7	6
9	5	1
4	3	8

And here are F_1 to F_9, the first 9 Fibonacci numbers: 1, 1, 2. 3, 5, 8, 13, 21, 34. Now replace the numbers in the square with the corresponding Fibonacci numbers.

1	13	8
34	5	1
3	2	21

It's no longer a magic square but multiply the numbers in each row and add.

$$(1 \times 13 \times 8) + (34 \times 5 \times 1) + (3 \times 2 \times 21) = 400$$

Do the same for the columns. And, of course:

$(1 \times 34 \times 3) + (13 \times 5 \times 2) + (8 \times 1 \times 21) = 400$

6

6 is not used so much these days. Those who were born before decimalization in February 1971 will remember the little silver sixpence commonly called the tanner. Older men may also remember the 'good old days' at school when they got 'six of the best'. In England this was six strokes of the cane on a delinquent's posterior. In Scotland, 'six of the best' was six strokes on the hand by an angry teacher wielding the tawse. Most of these old tyrants are now 'six feet under'.

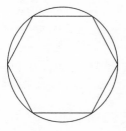

The figure above is a hexagon. Using compasses first draw a circle and, without changing the radius, mark a point on the circumference and then step around making a further 5 marks.

You will find that you finish exactly where you started off. You have divided the circumference into 6 equal arcs. With a straight edge, join up the marks to make a regular hexagon. Each angle is 120 degrees. The hexagon can be divided into 6 equilateral triangles.

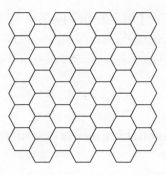

Hexagons fit together to make a honeycomb pattern. For the clever bees, this is the most efficient way of storing food.

It's Always 6

Back to the maths! Here is another little trick to demonstrate the magic of numbers. Write down any 3 consecutive numbers but the last one has to be divisible by 3 (see the '3' chapter for the divisibility rule). Add the 3 numbers, then add the digits of your answer and continue until you have a single digit.

$$10 + 11 + 12 = 33 \rightarrow 3 + 3$$
$$= 6$$
$$46 + 47 + 48 = 141 \rightarrow 1 + 4 + 1$$
$$= 6$$
$$1045 + 1046 + 1047 = 3138 \rightarrow 3 + 1 + 3 + 8$$
$$= 15 \rightarrow 1 + 5$$
$$= 6$$

Yes, you always end up with 6 and you never need too many steps to get there.

Divisibility by 6

That's easy. First check the number is even and then do the '3' test. At a glance, you can see that 1,000,000,002 is divisible by 6.

Hexagonal Numbers (Cornered)

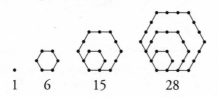

1   6   15   28

The above diagram shows the first 4 hexagonal numbers 1, 6, 15 and 28 and, of course, they go on forever. These are often called the cornered hexagonal numbers to distinguish them from the other series called the centred hexagonal numbers (see '37 and the Centred Hexagonal Numbers').

The formula for creating the cornered hexagonal numbers is a simple one:

n(2n – 1)

So let's check the 4th number, 4(8 – 1) = 4 × 7 = 28, and after calculating a few more, we have:

1, 6, 15, 28, 45, 66, 91, 120, 153, 190, 231, 276, 325, 378, 435, 496, 561 …

Here are the triangular numbers again:

1, 3, 6, 10, 15, 21, 28, 36, 45, 55, 66, 78, 91, 105, 120, 136, 153, 171, 190 ...

Every hexagonal number is also triangular but only every 2nd triangular number is hexagonal.

Sexy Primes

Tut-tut! What *are* you thinking about? There is nothing titillating about the sexy prime pairs. When you were reading about the hexagonal numbers, the '*hex*' part came from the Greek word for 6. The Latin word for the number 6 is '*sex*', as in 'sextuplets'. Perhaps it might have been preferable if mathematicians had called them the hexy prime pairs but then they wouldn't have enjoyed their little joke, would they?

You have a sexy prime pair when 2 adjacent prime numbers differ by 6. It is thought, but not yet proved, that there is an infinite number of them along the number line.

Some examples you already know in the Sieve of Eratosthenes, discussed earlier, are (23 and 29), (31 and 37), (53 and 59), (61 and 67), (73 and 79) and (83 and 89).

There are another 40 or so under 500 but there are 2 spectacular sexy prime pairs in Harry L. Nelson's magic square, also discussed earlier. They are (1,480,028,153 and 1,480,028,159) and (1,480,028,183 and 1,480,028,189).

Getting Even More Sexy

There are also sexy prime triplets of the form, (p, p + 6, p + 12), such as (47, 53, 59), and sexy prime quadruplets of the form, (p, p + 6, p + 12, p + 18), such as (61, 67, 73, 79). Getting more exciting all the time, what about sexy prime quintuplets? Alas, there is only one set on the whole number line and that is (5, 11, 17, 23, 29). Any starter other than 5 will have p + 6, p + 12, p + 18 or p + 24 – divisible by 5 and therefore not prime.

Perfect Numbers

6 is the first perfect number. Perfect? Really? Free from all flaws? Well, perhaps not, but mathematically speaking, it is faultless.

A perfect number is of the form σ(n) = n. The sign σ is the lower case of the Greek letter 'sigma' and is used by mathematicians when they want to say, 'the sum of'. So the little bit of apparent gobbledegook is a neat way of saying that a number is perfect when it is equal to the sum of its divisors (not including the number itself).

The divisors of 6 are 1, 2 and 3, which add up to 6. If you are underwhelmed by this, larger numbers are much more dramatic. The divisors of 28 are 1, 2, 4, 7 and 14, and these numbers add to 28. The next 2 perfect numbers are 496 with divisors 1, 2, 4, 8, 16, 31, 62, 124 and 248; and 8128 with divisors 1, 2, 4, 8, 16, 32, 64, 127, 254, 508, 1016, 2032 and 4064. Go on! You want to check it. Yes, it adds up to 8128.

After that, the perfect numbers get big-headed. The next one – the 5th one – is a long way off and very big. It is 33,550,336 and, amazingly, it was recorded by an unknown mathematician in the Middle Ages. It is not known if he (or she?) wrote down all the divisors.

Perfect numbers are closely linked to the Mersenne numbers and all of them are of the form $2^{p-1}(2^p - 1)$, but the bit in the bracket must not be merely a Mersenne number. It must be a Mersenne *prime* number. So $2^2 (2^3 - 1) = 4 \times 7 = 28$ is a perfect number because $2^3 - 1 = 7$, which is prime. $2^4(2^5 - 1) = 16 \times 31 = 496$ is perfect because the bit in the bracket is prime. Likewise, $2^6(2^7 - 1) = 8128$ is perfect because $(2^7 - 1) = 127$, which is prime.

Also perfect is $2^{12}(2^{13} - 1) = 4096(8192 - 1) = 4096 \times 8191 = 33{,}550{,}336$, the 5th perfect number above. This is because 8191 is prime. But $2^{10}(2^{11} - 1) = 1024 \times 2047 = 2{,}096{,}128$ is *not* perfect because 2047 is divisible by 23.

So all perfect numbers are of the form $2^{p-1}(2^p - 1)$ but not every number created by this formula is necessarily perfect. (Back to the 'all strawberries are red but not all red berries are strawberries' principle.)

Like the Mersenne prime numbers, the perfect numbers get very big, very quickly. The 10th one has 54 digits and you don't want to know about how big they get after that.

Each new possibility has to have its divisors checked and then added. With a computer, this may not be such a mammoth task as it was before, but let's take our hats off to the unknown medieval mathematician who discovered the 5th perfect number without any calculating aids whatsoever.

You may have noticed that 6, 28 and 496 are on the list of hexagonal numbers but so are all the other perfect numbers. *Every perfect number is hexagonal and therefore also triangular.*

There is a simple formula for testing if a number is hexagonal:

$$n = [\sqrt{(8x + 1)} + 1] \div 4$$

If n is a whole number, then x is a hexagonal number.

Let's test the perfect number 8128:

$$n = [\sqrt{(8 \times 8128 + 1)} + 1] \div 4$$
$$= [\sqrt{(65{,}024 + 1)} + 1] \div 4$$
$$= [\sqrt{(65{,}025)} + 1] \div 4$$
$$= (255 + 1) \div 4$$
$$= 256 \div 4 = 64$$

A nice, satisfying whole number.

Therefore the perfect number 8128 is also the 64th hexagonal number.

If you have a suitable calculator, go and prove that the 5th perfect number is hexagonal and which one along the line it is. *But remember that although every perfect number is hexagonal, every hexagonal number is not perfect.*

Mathematicians have also noticed that every perfect number, except 6, when divided by 9 has a remainder of 1.

Check the next 4:

$$28 \div 9 = 3 \text{ r } 1$$
$$496 \div 9 = 55 \text{ r } 1$$
$$8128 \div 9 = 903 \text{ r } 1$$
$$33{,}550{,}336 \div 9 = 3{,}727{,}815 \text{ r } 1$$

It seems that perfect numbers are always even and end with 6 or 8. No odd perfect number has ever been found but no one has been able to prove that every perfect number *must* be even.

Nicomachus was very fond of perfect numbers, particularly the first one. He said in his text, *Introducto Arithmetica*, 'God created all things in six days, because the number six is perfect.'

7

For the Pythagoreans, 7 was sacred and mystical and was another of their special numbers.

All the seven wonders of the ancient world have been destroyed by fire or earthquake except the Great Pyramid of Gisa, which was built nearly 5,000 years ago with awesome mathematical precision.

Also from the ancient world, we have the seven deadly sins, the seven hills of Rome, the seven ages in the life of man and the magical rainbow with its seven colours.

The month September comes from the Latin '*septem*', meaning 'seven'. September is, of course, our ninth month but the Romans added two extra months in commemoration of their famous leaders, Julius Caesar and Caesar Augustus. This also puts October, November and December (from the Latin, *octo*, *novem*, *decem*) out of sync.

In the twenty-first century, J. K. Rowling brought back the mysticism of 7 in her seven Harry Potter books. There were seven floors at Hogwarts, seven players in a Quidditch team, seven Weasley children and many other references to the number. Septima Vector was Harry's professor of arithmancy (the study of the magical properties of number). She issued a great deal of difficult homework, which only Hermione Granger enjoyed.

The author's father was proud to be the seventh son of a seventh son. Although he may have been the best dad in the world, unfortunately he had no supernatural powers.

Number Tricks Involving 7

Ask your friend to choose any number and add 9 to it. Multiply by 2, subtract 4 and divide by 2. Subtract the number they first chose. The answer is always 7. For example, start with 10.

$$10 + 9 = 19$$
$$19 \times 2 = 38$$
$$38 - 4 = 34$$
$$34 \div 2 = 17$$
$$17 - 10 = 7$$

If you can handle a little bit of easy algebra, this will explain why; x is your number, so you have:

$$x + 9 \rightarrow (2x + 18 - 4) \div 2 \rightarrow x + 7 \rightarrow 7$$

In another number trick, allow your friend to borrow your pocket calculator. Ask them to write down their age, multiply it by 37, then by 13, then by 3 and, just for luck, multiply once more by 7. For example:

$$49 \times 37 \times 13 \times 3 \times 7 = 494949$$

There it is – their age in triplicate.

Divisibility by 7

Double the last digit and subtract it from the remainder of the number. Keep doing this until you do or do not have a number on the 7 times table. Examples:

1. Does 616 divide by 7?

   61 − 12 = 49 (yes)

   So 616 is divisible by 7.

2. Does 19,792 divide by 7?

   1979 − 4 = 1975
   197 − 10 = 187
   18 − 14 = 4

   So 19,792 is not divisible by 7.

3. Does 39,361 divide by 7?

   3936 − 2 = 3934
   393 − 8 = 385
   38 − 10 = 28

   So 39,361 is divisible by 7.

7 is important because it has the most fascinating reciprocal in the whole number line:

$1 \div 7 = 0.142857142857 \ldots$

There is no rule to say that you must read this book in order so if you can't wait to see why this is, off you go to 142,857.

At 6s and 7s

When things are 'at sixes and sevens', it means they are confused or in disarray.

These 6s and 7s below are very much the opposite and are very tidy indeed.

$$6 \times 7 = 42$$
$$66 \times 67 = 4422$$
$$666 \times 667 = 444,222$$
$$6666 \times 6667 = 44,442,222$$
$$66666 \times 66667 = 4,444,422,222$$

And keep going for as long as you like.

Heptagonal Numbers

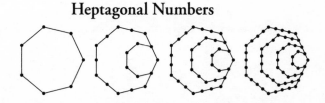

The Greek word for 7 is '*hepta*' so the second figure above is a heptagon. As we know already, 1 likes to be in on everything so, counting spots, the first 5 heptagonal numbers are: 1, 7, 18, 34 and 55 but the formula ½n(5n – 3) will help you calculate as many as you want.

The seventh heptagonal number will be ½ × 7 × (5 × 7 – 3) = ½ × 7 × 32 = 112. So writing out a string of them from the beginning, we get:

1, 7, 18, 34, 55, 81, 112, 148, 189, 235 …

The difference between two successive numbers is given by the formula $5n - 4$ where n is the larger number. Test it for the difference between the ninth and tenth numbers.

$5n - 4 = 5 \times 10 - 4 = 46$ (235 − 189 = 46, correct)

Do you notice that they go odd, odd, even, even, odd, odd, even, even …? They go on like this forever.

Here are those triangular numbers again:

1, 3, <u>6</u>, 10, 15, 21, 28, <u>36</u>, 45, 55, 66, 78, <u>91</u>, 105, 120, 136, 153, <u>171</u>, 190, 210, 231, 253 …

So you can see that 5 times a heptagonal number plus 1 is a triangular number. Check from the beginning of the pentagonal numbers list:

$5 \times 1 + 1 = 6$
$5 \times 7 + 1 = 36$
$5 \times 18 + 1 = 91$
$5 \times 34 + 1 = 171$

Yes, it checks and, after 6, this applies to every 5th triangular number along the line.

Happy Numbers

Did you know that there are lots of happy numbers along the number line? Apart from trivial old 1, which, as we know, likes to be at every party, 7, 10, 13, 19, 23, 28, 31, 32, 44, 49, 68, 70, 79, 82, 86, 91, 94 and 97 are all deliriously happy, to list only those less than 100.

To see if a number has happy status is not too lengthy a procedure. Start with any number, replace it with the sum of the square of its digits and repeat this process; if the number stops at 1, you must have started with a happy number. If the number gets itself into an endless loop going round and round and round, you have an unhappy number or, better still, a very dizzy number.

Let's test 7 to confirm that it is happy:

$$7 \rightarrow 7^2 \rightarrow 49 \rightarrow 4^2 + 9^2 \rightarrow 97 \rightarrow 9^2 + 7^2 \rightarrow 130 \rightarrow 1^2 + 3^2 + 0^2 \rightarrow 10 \rightarrow 1^2 + 0^2 \rightarrow 1$$

And let's test 82:

$$82 \rightarrow 8^2 + 2^2 \rightarrow 68 \rightarrow 36 + 64 \rightarrow 100 \rightarrow 1^2 \rightarrow 1$$

78,999 looks rather off-putting. Want to try it?

$$78,999 \to 7^2 + 8^2 + 9^2 + 9^2 + 9^2 = 356 \to 3^2 + 5^2 + 6^2 =$$
$$70 \to 7^2 + 0^2 = 49 \to 4^2 + 9^2 = 97 \to 9^2 + 7^2 = 130 \to 1^2 +$$
$$3^2 + 0^2 = 10 \to 1^2 + 0^2 \to 1$$

There now! That wasn't too bad, was it? It has been conjectured that every happy number needs a maximum of 7 steps to get to its target.

But what about, say, 42? Without detailing every step as before:

$$42 \to 20 \to 4 \to 16 \to 37 \to 58 \to 89 \to 145 \to 42$$

We are back at 42 again so 42 is an unhappy, dizzy number.

Once you have determined that a number is happy, all the intermediate numbers in the calculation are also happy.

With any happy number, say, 79, any other arrangement of the digits, for example 907, 7009, 9700, etc. will also be happy numbers and likewise for the unhappy numbers. So the unique combination of happy numbers under 100 (the rest are just rearrangements and/or insertions of zeros anywhere in the number) is: 1, 7, 13, 19, 23, 28, 44, 49, 68, 79.

In an episode of *Doctor Who*, a sequence of happy prime numbers (313, 331, 367, 379) is used as a code to unlock a sealed door on a spaceship that is about to collide with the sun. Only the Doctor himself knew about happy numbers and he saved the whole crew from disaster.

8

At one time, eight pints of beer was thought to be a reasonable amount to drink. So someone who was 'one over the eight' was probably only slightly drunk.

The Chinese super-lucky number is 8. In Mandarin the word for 8 sounds like the word for 'prosperity'. They want 8 in their car registration numbers, their house numbers, their phone numbers, everywhere in their lives. The United Airline flight between Beijing and San Francisco is UA 888. The opening ceremony of the 2008 Olympic Games in Beijing was on 8 August. And Sichuan Airways bought the phone number 88888888 for an enormous sum of money. However, it was worth every yuan. With all that luck, anyone would fly with that company to the moon and back with complete confidence.

Number Patterns with 8

8 appears in several number patterns. This one involves triangular numbers and square numbers:

$$8 \times 1 + 1 = 3^2$$
$$8 \times 3 + 1 = 5^2$$
$$8 \times 6 + 1 = 7^2$$
$$8 \times 10 + 1 = 9^2$$
$$8 \times 15 + 1 = 11^2$$
$$8 \times 21 + 1 = 13^2$$

And on and on forever.

Another pattern involving 8:

$1 \times 8 + 1 = 9$
$12 \times 8 + 2 = 98$
$123 \times 8 + 3 = 987$
$1234 \times 8 + 4 = 9876$
$12345 \times 8 + 5 = 98765$
$123456 \times 8 + 6 = 987654$
$1234567 \times 8 + 7 = 9876543$
$12345678 \times 8 + 8 = 98765432$
$123456789 \times 8 + 9 = 987654321$

And one more:

$9 \times 9 + 7 = 88$
$98 \times 9 + 6 = 888$
$987 \times 9 + 5 = 8888$
$9876 \times 9 + 4 = 88888$
$98765 \times 9 + 3 = 888888$
$987654 \times 9 + 2 = 8888888$
$9876543 \times 9 + 1 = 88888888$
$98765432 \times 9 + 0 = 888888888$

Also:

$8^3 = 512$ and $5 + 1 + 2 = 8$

Divisibility by 8

A number is divisible by 8 if the number made by the last 3 digits is divisible by 8. So 34,587,<u>592</u> is divisible by 8; 453,<u>882</u> is not.

The Octagonal Numbers

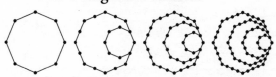

Lastly, we will have a look at the octagonal numbers. (There are nonagonal numbers and decagonal numbers, you know ... but maybe enough is enough.)

Counting the spots again, the first 5 are 1, 8, 21, 40 and 65. The formula is $3n^2 - 2n$, so calculate the 12th one.

$3 \times 12 \times 12 - 2 \times 12 = 432 - 24 = 408$

Now write out the list as far as you wish to go:

1, 8, 21, 40, 65, 96, 133, 176, 225, 280, 341, 408 ...

Notice that the numbers alternate odd, even, odd, even ... and, as you might have come to expect, they keep doing this forever.

A Review of Polygonal Numbers

shape	formula	\multicolumn{9}{c}{n =}								
		1	2	3	4	5	6	7	8	9
triangle	$\frac{1}{2}n(n+1)$	1	3	6	10	15	21	28	36	45
square	n^2	1	4	9	16	25	36	49	64	81
pentagon	$\frac{1}{2}n(3n-1)$	1	5	12	22	35	51	70	92	117
hexagon	$n(2n-1)$	1	6	15	28	45	66	91	120	153
heptagon	$\frac{1}{2}n(5n-3)$	1	7	18	34	55	81	112	148	189
octagon	$3n^2-2n$	1	8	21	40	65	96	133	176	225

So what do you notice?

a) In the n = 3 column, all the numbers are divisible by 3.
b) In the n = 4 column, the numbers increase by 6 each time.
c) In the n = 5 column, the numbers are divisible by 5 and increase by 10 each time.
d) In the n = 6 column, the numbers increase by 15 each time
e) In the n = 7 column, the numbers divide by 7 and increase by 21 each time.
f) In the n = 8 and n = 9 columns, there are increases of 28 and 36, and so on.

You can see the pattern of increases is 1, 3, 6, 10, 15, 21, 28, 36 and these are the triangular numbers. If another column for n = 10 had been added, the vertical increase would have been 45. Go and check.

Now from the table take any square block. For example:

34   55
40   65

Cross multiply and subtract, getting 2210 – 2200 = 10, then look at the number at the top of the '34' column. It's 10, isn't it? Take another block:

148   189
176   225

Do the same: 148 × 225 – 176 × 189 = 333,000 – 33,264 = 36. What do you see at the top of the '148' column?

Magic, isn't it?

Factorial!

Sometimes, you see the symbol ! in maths. It's exactly the same symbol as in written English. 'He's dead!' means, 'My goodness, he has died. We never expected him to die. How could this have happened?' So in maths, does '5!' mean '5? Really? Wow, is it that many?'

No, sorry to disappoint you. There is no drama in a mathematical ! The symbol ! is called factorial. Some say factorial 5, others say 5 factorial. 5! simply means $5 \times 4 \times 3 \times 2 \times 1$, which equals 120. Likewise 8! saves you a lot of writing and the answer is 40,320.

1! = 1 but 0! = 1 also. We only know this because the greatest mathematicians have decreed that this is so.

Number Patterns with Powers and Factorials

Write down some powers of 3 and in the next line subtract in pairs. Continue until all the numbers are the same.

$$1^3 \quad 2^3 \quad 3^3 \quad 4^3 \quad 5^3 \quad 6^3 \quad 7^3 \quad 8^3 \rightarrow \text{more if you want}$$

$$\begin{array}{ccccccccc}
1 & & 8 & & 27 & & 64 & & 125 & & 216 & & 343 & & 512 \\
& 7 & & 19 & & 37 & & 61 & & 91 & & 127 & & 169 \\
& & 12 & & 18 & & 24 & & 30 & & 36 & & 42 \\
& & & 6 & & 6 & & 6 & & 6 & & 6
\end{array}$$

And $6 = 3 \times 2 \times 1 = 3!$

Try it again with powers of 5:

$$1^5 \quad 2^5 \quad 3^5 \quad 4^5 \quad 5^5 \quad 6^5 \quad 7^5 \rightarrow \text{more if you want}$$

```
  1    32   243  1024 3125 7776 1680
     31   211   781  2101 4651  903
        180  570 1320 2550 4320
           390  750 1230 1830
              360  480  600
                 120  120
```

And 120 is 5!

Try it yourself for powers of 4. Guess what you'll end up with! It will work for higher powers but the numbers get rather unmanageable.

There Are N! Ways of Arranging N Objects

Suppose an artist wanted to arrange his 3 favourite paintings on his wall. They were all still-life paintings: one of a bowl of apples, one of a bunch of bananas and the other a box of freshly picked cherries. He just couldn't make up his mind in which order to hang them. He had 6 choices, ABC, ACB, BAC, BCA, CAB and CBA, because $3! = 3 \times 2 \times 1 = 6$.

If he had added his recent work of a date palm tree painted in Tunisia, the poor man would have had $4! = 24$ arrangements to choose from.

Just to confirm this simple calculation, here are all his options:

ABCD	BACD	CABD	DABC
ABDC	BADC	CADB	DACB
ACBD	BCAD	CBAD	DBAC
ACDB	BCDA	CBDA	DBCA
ADBC	BDAC	CDAB	DCAB
ADCB	BDCA	CDBA	DCBA

Factorial Primes

$n! \pm 1$

A factorial prime number is a prime number that is one more or one less than a factorial. The first 10 factorials excluding 0! are:

$1! = 1$
$2! = 2$
$3! = 6$
$4! = 24$
$5! = 120$
$6! = 720$
$7! = 5040$
$8! = 40,320$
$9! = 362,880$
$10! = 3,628,800$

10! is already in the millions so we won't go further than that. All factorial numbers are even except 1! So to look for primes we have to test both sides of the factorial.

1!, 2! and 3! generate primes 2, 3, and 7 by adding 1; and 3! (again), 4!, 6! and 7! generate 5, 23, 719 and 5039 by subtracting 1. The others in the above list fail. So, once again, we do not have a formula that only produces primes.

However, as the factorials get bigger and bigger, huge prime numbers appear from time to time, the largest at the time of writing being 150,209! + 1 with 712,355 digits and, yes, it has been tested.

Factorial Fun

$3! - 2! + 1! = 5$
$4! - 3! + 2! - 1! = 19$
$5! - 4! + 3! - 2! + 1! = 101$
$6! - 5! + 4! - 3! + 2! - 1! = 619$
$7! - 6! + 5! - 4! + 3! - 2! + 1! = 4421$

Keep going for a few more. The answers above are all prime numbers – so far.

$(n!)! = n!(n!-1)!$

At the beginning of this chapter, this would have been gobbledygook, wouldn't it? But now, it's easy. Let's confirm that the statement is correct by letting n = 3. Take the left-hand side and right-hand side one at a time.

LHS = (3!)!   RHS = 3! (3! –1)!

$$\text{LHS} = (3!)! \qquad\qquad \text{RHS} = 3! \, (3! - 1)!$$
$$= (3 \times 2 \times 1)! \qquad\qquad = 6 \, (6 - 1)!$$
$$= 6! \qquad\qquad\qquad\quad = 6 \times 5!$$
$$= 6 \times 5 \times 4 \times 3 \times 2 \times 1 \qquad = 6 \times 120$$
$$= 720 \qquad\qquad\qquad\quad = 720$$

Try it for another value of n.

Subfactorials !n

Suppose the postman has 2 letters to deliver to 2 houses. How many ways can he get it wrong? He could put letter A into box A, in which case he has got it correct, or he could put letter B into A's box, in which case he's got it wrong. The possibilities are AB and BA. So with 2 letters, he can get it completely wrong *once*. So !2 (subfactorial 2) = 1.

Now with 3 letters and 3 addresses, this could happen:

ABC	BAC
ACB	~~CAB~~
~~BCA~~	CBA

There are only 2 occasions when he gets it *completely* wrong and *none* of the letters go to the correct house. So the number of ways to deliver 3 letters to 3 addresses is 3! = 3 × 2 × 1 = 6.

And the number of ways to get it totally deranged is !3 = 2. (Note that mathematicians are using 'deranged' in the literal sense.)

Now what is this careless postman likely to do with 4 letters to 4 houses?

ABCD	BACD	CABD	~~DABC~~
ABDC	~~BADC~~	~~CADB~~	DACB
ACBD	BCAD	CBAD	DBAC
ACDB	~~BCDA~~	CBDA	DBCA
ADBC	~~BDAC~~	~~CDAB~~	~~DCAB~~
ADCB	BDCA	~~CDBA~~	~~DCBA~~

Although there are occasions when he gets it *mostly* wrong, like BDCA, at least he puts C into the correct box even if A, B and D are in the wrong places. But there are 9 occasions when the letters are *all* in the wrong place. So 4! = 24 and !4 = 9.

It is impossible to get 1 letter wrong if there is only 1 box to put it in. So we can start a list of subfactorial numbers with !1 = 0. To avoid any further ABCD … tables there is a formula to calculate any subfactorial !n as long as you know the 2 subfactorials before it. On our subfactorial list so far we have the first 4: 0, 1, 2, 9. Here is the formula to get as many as you fancy, starting with !5:

$$!n = (n-1)[!(n-1) + !(n-2)]$$
$$!5 = 4[!4 + !3]$$
$$!5 = 4(9 + 2)$$
$$= 44$$

And for !6:

$$!6 = 5(!5 + !4)$$
$$= 5(44 + 9)$$
$$= 265$$

The !n list continues and gets very big, very quickly: 0, 1, 2, 9, 44, 265, 1854, 14,833, 133,496 …

Subfactorials adapt for many other situations – not just careless posties.

148,349

Would you believe it? Some brilliant mathematician has discovered that 148,349 equals the sum of its subfactorials. (To avoid confusion, the exclamation mark that the statement deserves has been omitted.)

148,349 = !1 + !4 + !8 + !3 + !4 + !9

You have the list of subfactorials above if you want to check it yourself.

$$\infty$$

The figure of 8 is traced out on the ice by a skater who could go round and round the shape ad infinitum. The same shape turned on its side is ∞, the mathematical symbol for infinity. ∞ is at the end of forever. It is not an actual number but the concept of something so large that it is limitless and beyond our reach.

9

Lady Jane Grey had the shortest reign in English history. In 1553, she was on the throne for only nine days. Soon afterwards, she was beheaded.

Cats are said to have nine lives. If they fall from a tree, they usually land neatly on well-padded paws. Office workers have a 9 to 5 job, we go to a wedding 'dressed up to the nines' and 9 times out of 10, the train is late.

9 is a well-behaved number. It follows all the rules. The digits of the 9 times table numbers add up to 9 and the rule for divisibility by 9 is equally simple.

Divisibility by 9

Add the digits. If the total is a number on the 9 times table, the original number is divisible by 9. 121212 must divide by 9 exactly and so must 531486.

The reciprocal of 9, 1 over 9, is 0.11111111 … and the reciprocal of its next-door neighbour at No. 11 is 0.09090909 …

A Number Puzzle Where You Always End Up with 9

Take any number, say 25 and add the digits, giving you 7. Take 7 from 25, giving 18; add the digits and you get 9.

Getting a bit more ambitious:

$165 \rightarrow 1 + 6 + 5 = 12$. $165 - 12 = 153 \rightarrow 1 + 5 + 3 = 9$

And once again:

$1728 \rightarrow 1 + 7 + 2 + 8 = 18$. $1728 - 18$
$= 1710 \rightarrow 1 + 7 + 1 + 0 = 9$

11

Hold on! 10 is missing. Well, 10 is not especially interesting. All of us can add 10, subtract 10, multiply and divide by 10, and we all know about the ten commandments including the one that commands us not to covet our neighbour's ox. So let's move on.

11 is far more important – well, it certainly was in 1752. At the end of the sixteenth century, most of Europe changed to the more accurate Gregorian calendar. Britain, however, under the rule of Queen Elizabeth, refused to adopt the new popish calendar. Elizabeth had not forgotten all the trouble her father, King Henry VIII, had with the Roman Catholic Church when he wanted to marry her mother, Anne Boleyn. However, by 1752, Britain had fallen behind the rest of Europe by eleven days. So to catch up, George II ordered that the Gregorian calendar should be adopted and that Wednesday, 2 September should be followed by Monday, 14 September.

This caused havoc. Many people had no birthday that year. The twin born two minutes after midnight was eleven days younger than his brother. Some thought that their lives had been shortened by eleven days and, what was worse, they had lost eleven days' wages. They marched in the streets protesting 'Give us back our eleven days.'

The Gregorian calendar is inaccurate by just one extra day in 3,323 years. So, counting from 1752, another small adjustment will have to be made in the year 5075.

If anyone is around at that time, please arrange to have the extra day in June, preferably on a sunny Saturday.

The Repunits

11 is a repunit. 'Repunit' means unity or the digit 1 repeated. Since the word does not appear in the *Oxford English Dictionary*, you won't know how to pronounce it. You say, 're-pew-nit'. The 'pun' bit does not rhyme with 'fun' although the repunits are fun – well, mathematical fun, anyway.

At a board meeting of repunits, it was decided unanimously that 1 could not become a member of the exclusive repunit club as it only had one leg. However, 1 is a member of so many other clubs including the Happy Number Club, where it is the very important founder member who decides which other numbers may join the club. And in the famous Fibonacci sequence, 1 is not only the first number, it is also the second!

So 'legs 11' is the smallest repunit and the others, 111, 1111, 11111... go on forever. 111 and every 3rd one after that divides by 3; repunits with an even number of legs divide by 11. Some of them are chock-a-block with prime factors. The one with 30 legs, $R_{30} = 3 \times 7 \times 11 \times 13 \times 31 \times 37 \times 41 \times 211 \times 241 \times 271 \times 2161 \times 9091 \times 2,906,161$. (Don't worry, that last factor *is* prime. Greater brains than ours have proved it.) Prime number repunits are very rare. Little legs 11 is a prime, of course, but the only other primes in the first 100 repunits (yes, that's all the way up to a 100-legged monster) are R_{19} and R_{23}. Beyond that, there is only one more known repunit prime and it has 317 legs. To write it out on A4 paper, with single spacing between the units, takes over six lines.

You don't really need a formula for repunits but $(10^n - 1) \div 9$ fits them all.

When squared, the repunits make pretty patterns:

$11 \times 11 = 121$
$111 \times 111 = 12321$
$1111 \times 1111 = 1234321$

And up and up you can go to:

$111111111 \times 111111111 = 12345678987654321$

It doesn't work after that because a 10-legged repunit multiplied by another 10-legged repunit would have 10 in the middle of the answer, necessitating a carried 1, and the pattern becomes very messy. But 13 legs × 13 legs is nicely symmetrical and equals:

12345678900987654321

But that isn't all. Continuing from the above pattern:

$121 = 22 \times 22 \div (1 + 2 + 1)$
$12321 = 333 \times 333 \div (1 + 2 + 3 + 2 + 1)$
$1234321 = 4444 \times 4444 \div (1 + 2 + 3 + 4 + 3 + 2 + 1)$

... all the way to the monster:

12345678987654321
$= 999999999 \times 999999999 \div (1 + 2 + 3 + 4 + 5 + 6 + 7 + 8 + 9 + 8 + 7 + 6 + 5 + 4 + 3 + 2 + 1)$

Now, if you add the digits of the above squared numbers of repunits from 2 digits long to 9 digits long (121, 12321, etc.), you get 4, 9, 16, 25 … yes, the squares of 2 to 9.

Here is another pattern just to show you how well-behaved numbers can be:

$1 \times 9 + 2 = 11$
$12 \times 9 + 3 = 111$
$123 \times 9 + 4 = 1111$
$1234 \times 9 + 5 = 11111$

And so on, but not forever as the pattern dies after $123456789 \times 9 + 10 = 1111111111$.

More Magic with the Repunits

$11 = 6^2 - 5^2$
$1111 = 56^2 - 45^2$
$111111 = 556^2 - 445^2$
$11111111 = 5556^2 - 4445^2$

And so on.

There is a similar pattern for:

$33 = 7^2 - 4^2$
$3333 = 67^2 - 34^2$
$333333 = 667^2 - 334^2$

And so on. (The general name for a number with a repeating digit such as 3333 is a 'repdigit'.)

One more for 8s and 3s:

$$55 = 8^2 - 3^2$$
$$5555 = 78^2 - 23^2$$
$$555555 = 778^2 - 223^2$$

Try it yourself for 9s and 2s. What will be the repeated digit?

A Quick Way to Multiply by 11

By the following method, multiplying even a large number by 11 only takes a few seconds. For example: 724361 × 11.

Write down the first and last digits with a space in between:

7   1

Now add the digits in pairs from right to left:

$$1 + 6 = 7$$
$$6 + 3 = 9$$
$$3 + 4 = 7$$
$$4 + 2 = 6$$
$$2 + 7 = 9$$

Now fill in 96797 in the space.

Answer: 7967971

You have to work from right to left because if the pair of digits adds to more than 9, for example $6 + 8 = 14$, you have to write down the 4 and carry 1.

We can make up a little number trick using the easy way to multiply 11s. Choose any number at all whose adjacent digits do not add up to more than 9. For example, 4516.

$4516 \times 11 = 49,676$

Reverse the digits.

$6154 \times 11 = 67,694$

Divisibility by 11

A number is divisible by 11 if the sum of the 1st, 3rd, 5th and so on digits equals the sum of the second, fourth, sixth and so on digits or if the difference is a multiple of 11. So 12,320, 1573 and 979 are divisible by 11 but 8395 is not.

12

$12^2 = 144$, and reversing all digits, $21^2 = 441$. Of course, that's interesting but not very interesting. However, before dismissing 12, let's have fun with another number system.

The Duodecimal System

The headings for the decimal system that we all use are:

←1000s   100s   10s   1s

The headings for the duodecimal system ('*duodecim*' is Latin for twelve) are:

←1728s   144s   12s   1s

For the decimal system, we have 10 digits: 0, 1 , 2, 3, 4, 5, 6, 7, 8, 9. But the duodecimal system needs 12 digits so 2 extra have to be invented. Some mathematicians use A and B for the extra digits but T and E make more sense because they stand for 10s and 11s. So you have the 12 digits, 0, 1, 2, 3, 4, 5, 6, 7, 8, 9, T and E.

Nothing exciting starts for a while. The numbers 0 to 9 are the same in both bases but 10_{10} and 11_{10} are T_{12} and E_{12}. It seems strange but it isn't difficult. Write down the duodecimal headings again and you are going to change the following base 10 numbers to base 12:

60, 122, 154, 299, 298, 500, 1440, 1739

OK, think in bundles of 12s:

←1728s   144s   12s   1s

1. 60_{10} is 5 bundles of 12 so 5 goes in the 12s column and 0 in the 1s column:

    $60_{10} = 50_{12}$

2. 122_{10} is 10 bundles of 12 plus 2 extra so T goes in the 12s column and 2 in the 1s:

    $122_{10} = T2_{12}$

3. 154_{10} is 1 bundle of 144s , 0 bundles of 12s and a 10 extra:

    $154_{10} = 10T_{12}$

4. 299_{10} is 2 bundles of 144s, 0 bundles of 12s and 11 extra:

    $299_{10} = 20E_{12}$

5. 298_{10} is 2 bundles of 144s, 0 bundles of 12s and 10 extra:

    $298_{10} = 20T_{12}$

6. 500_{10} is 3 bundles of 144s, 5 bundles of 12s and 8 extra:

$$500_{10} = 358_{12}$$

7. 1440_{10} is 10 bundles of 144s exactly. Put T in the 144s column and fill the other headings with 0:

$$1440_{10} = T00_{12}$$

8. 1739_{10} is 1 bundle of 1728s with 11 extra and 11_{10} is E_{12}.

$$1739_{10} = 100E_{12}$$

If you've got a bigger number, just add another heading. The next one is 20,736.

It's even easier going from duodecimal to decimal. You need the headings again:

1728s   144s   12s   1s

1. TTE_{12} is $(10 \times 144) + (10 \times 12) + (11 \times 1) = 1571_{10}$
2. $10T0_{12}$ is $(1 \times 1728) + 0 + (10 \times 12) + 0 = 1848_{10}$

Try 111_{12} and $T1T_{12}$ yourself. The answers are 157_{10} and 1462_{10}.

You could even try a little adding sum. When is TT + TT equal to 198? Answer, when it is added in base 12 (T + T = 20, that's 8 carry 1. 1 + T + T = 21, that's 9, carry 1).

The Dozenal Society

The Dozenal Society of Great Britain was founded in 1959 and flourishes still. The equivalent society in the United States started in 1944 and both have a common aim. They maintain that the decimal system that we all use is not fit for purpose and they want to replace it with the duodecimal system.

They remind us that the year has 12 lunar months and the clock has 12 hours. There are 12 inches in a foot and men like to buy a dozen red roses for their sweethearts. They point out that 12's beautiful divisors, 1, 2, 3, 4 and 6, are just perfect for packing and packaging. The fractions most commonly used are ½, ⅓, ⅔, ¼ and ¾. Whoever would cut a cake into ⅕ths or ⅒ths?

Will we ever change? The mind boggles at the mere thought of it. Are the Dozenal Society members all crackpots? One cannot possibly comment.

Abundant Numbers

The maths speak for abundant numbers is $\sigma(n) > n$. Remember that σ is the Greek lowercase letter sigma and mathematicians use sigma for 'the sum of'. If a number is abundant, it means that the sum of its divisors is more than the number itself. 12 is the first abundant number. Its divisors are 1, 2, 3, 4 and 6, which add up to 16.

20 is also abundant with its divisors 1, 2, 4, 5 and 10.

All multiples of the perfect numbers are abundant. Back in the 'Perfect Numbers' section, you learned that 28 is a perfect number. So 56 must be abundant. The divisors of 56, which are 1, 2, 4, 7, 8, 14 and 28, add up to 64 so 56 has an 'abundancy' of 8. (Mathematicians love to use words that are not in the *Oxford English Dictionary*.)

The abundant numbers aren't all that abundant. There are 21 of them under 100:

12, 18, 20, 24, 30, 36, 40, 42, 48, 54, 56, 60, 66, 70, 72, 78, 80, 84, 88, 90, 96

These are all even but there are odd abundant numbers further along the number line. The first odd one is 945, whose factors are 1, 3, 5, 7, 9, 15, 21, 27, 35, 45, 63, 105, 135, 189 and 315, giving a total of 975.

Old Nicomachus (remember him?) in his *Introducto Arithmetica* didn't like abundant numbers. He described them as deformed animals with too many limbs.

Deficient Numbers

Deficient numbers are the opposite of abundant numbers: $\sigma(n) < n$. The sum of the divisors is less than the number itself. Obviously all prime numbers are severely lacking as their only divisor is 1.

Apart from the perfect 6 and 28, the deficient numbers under 100 are all the numbers not on the abundant list.

Nicomachus did not have a good word to say about deficient numbers either. He said in his book that they were like hands with fingers missing and many other nasty things besides.

13

13 is our unlucky number although young people are not obsessively concerned about it so much these days. 13 was regarded as unlucky by the Romans, who thought it was the symbol of death.

There were thirteen men at the Last Supper. The twelve disciples sat with Jesus, and Judas, who betrayed him, was on his left side. Most hosts, therefore, try to avoid having a dinner party with thirteen people sitting at the table.

There is always a Friday the 13th if the first day of the month is a Sunday. It is considered particularly unlucky and is definitely not a day for getting married, doing a parachute jump or having major surgery.

13 has an interesting reciprocal. $1 \div 13 = 0.076923076923$ … the decimal goes on forever but repeats after 6 digits.

Take one block of 076923:

$76923 \times 1 =$ 076923
$76923 \times 10 =$ 769230
$76923 \times 9 =$ 692307
$76923 \times 12 =$ 923076
$76923 \times 3 =$ 230769
$76923 \times 4 =$ 307692

Note that the '076923' sequence of digits is there both across and down, and each digit repeats on a south-west diagonal.

When 76923 is multiplied by another set of numbers, there is a different pattern:

$76923 \times 2 = 153846$
$76923 \times 7 = 538461$
$76923 \times 5 = 384615$
$76923 \times 11 = 846153$
$76923 \times 6 = 461538$
$76923 \times 8 = 615384$

Again, there is the across and down sequence of digits and diagonal rows of the same digit.

13 is the sum and difference of 2 consecutive squares and cubes:

$$2^2 + 3^2 = 13$$
$$7^2 - 6^2 = 13$$

And 13^2 is the difference of 2 consecutive cubes:

$$8^3 - 7^3 = 13^2$$

Growing Primes from 13

13
139
1399
13999
139991
1399913
13999133

But 139991333 is not a prime as it divides by 47.

13 has Palindromic Properties

A palindrome is any word, phrase or number that reads the same backwards as forwards, like 'madam' or 'nurses run' or 101.

$13 \times 62 = 26 \times 31$
$13 \times 93 = 39 \times 31$
$13 \times 13 = 169$ and $961 = 31 \times 31$

Divisibility by 13

Add 4 times the last digit to the rest of the number and continue until you can go no further.

For example: Is 3354 divisible by 13?

$335 + 16 = 351$
$35 + 4 = 39$

39 is divisible by 13, so 3354 is also divisible by 13.

Look at the reciprocal of 13 again, 0.076923 … Split it in half and add the 2 halves together: $076 + 923 = 999$. Remember that when you meet the reciprocals of other prime numbers later on.

An Anagram for 13

The letters in 'eleven plus two' can be rearranged to become 'twelve plus one'.

Create Art with 13

Draw a circle, any size you like, and mark 10 notches at equal distance around the circle starting at top right. (The angles at the centre of the circle will each be 360 degrees ÷ 10 = 36 degrees.) Number the notches from 0 to 9, again starting at top right. Now return to the 2 blocks of numbers found by calculating the reciprocal of 13 on the previous page. They are:

076923 and 153846

Just join the dots of the first block of numbers and again with the second block of numbers and sit back and enjoy the beautiful design you have created.

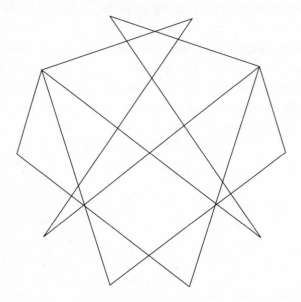

Emirps

13 has never been a popular number but recently it has found new fame for being the first emirp. An emirp (prime spelled backwards) is a prime number whose reversal is also prime. This does not include palindromic primes like 11 or 101.

The list starts 13, 17, 31, 37, 71, 73, 79, 97, 107, 113 … and on and on they go.

In Harry Nelson's spectacular magic square of 10-digit primes, only the first one is an emirp.

There are some very big emirps, but there's no fun in checking these. Here are six 4-digit primes that are very near to each other on the number line. They all have emirpy partners:

9721, 9749, 9769, 9781, 9787, 9791

From now on, many numbers won't get a special heading and some won't even be mentioned. Of course, we know from the tongue-in-cheek Interesting Numbers Paradox that all numbers are interesting. However, due to lack of space, only the very best numbers will be included from this point onwards.

16 and the Luhn Algorithm

Most credit cards and debit cards have 16 digits. The first digit on a Visa card is 4, on a Mastercard it is 5, and on American Express it's 3.

The following test for the validity of a bank card number was invented in 1954 by German IBM scientist Hans Peter Luhn, before most people even had debit or credit cards. Luhn's Algorithm is a simple bit of arithmetic designed not to guard against malicious attacks, but to protect against accidental errors. (You know the sort of thing – in the card number below you might write down 44317 instead of 44371, and the recipient asks you to repeat your number. Luhn's Algorithm spots that error immediately.)

This is how it works: write out the 16 digits of a bank card number, well spaced out, and underline the first one and every alternative one.

$$\underline{4} \quad 4 \quad \underline{3} \quad 7 \quad \underline{1} \quad 2 \quad \underline{1} \quad 4 \quad \underline{5} \quad 6 \quad \underline{1} \quad 8 \quad \underline{9} \quad 1 \quad \underline{7} \quad 3$$

Double all the underlined numbers:

$$8 \quad 6 \quad 2 \quad 2 \quad 10 \quad 2 \quad 18 \quad 14$$

Now add all the numbers that are *not* underlined and the doubled numbers, treating the 2-digit numbers as separate digits: 10 as 1 + 0 and 18 as 1 + 8. So you have:

4 + 7 + 2 + 4 + 6 + 8 + 1 + 3 + 8 + 6 + 2 + 2 + 1 + 0 + 2 + 1 + 8 + 1 + 4 = 70

The fact that 70 is divisible by 10 shows that the card is valid and no error has been made.

American Express cards have only 15 digits, such as:

3 $\underline{7}$ 8 $\underline{2}$ 8 $\underline{2}$ 2 $\underline{4}$ 6 $\underline{3}$ 1 $\underline{0}$ 0 $\underline{0}$ 5

This time, underline the 2nd digit from the left and every alternate one. Double all the underlined numbers and add them as before:

(1 + 4) + 4 + 4 + 8 + 6 + 0 + 0 = 27

Add on the non-underlined numbers:

3 + 8 + 8 + 2 + 6 + 1 + 0 + 5 = 33

The total of the 2 sets of numbers is 60, again a multiple of 10.

Diners Club cards have 14 digits and start with 300 or 305, such as the following:

$\underline{3}$ 0 $\underline{5}$ 6 $\underline{9}$ 3 $\underline{0}$ 9 $\underline{0}$ 2 $\underline{5}$ 9 $\underline{0}$ 4

Treat this number exactly like the 16-digit Visa or Mastercard one.

Doubling the underlined numbers and adding the non-underlined numbers gives:

$$6 + (1 + 0) + (1 + 8) + 0 + 0 + (1 + 0) + 0 + 0 + 6 + 3 + 9 + 2 + 9 + 4 = 50$$

It's another multiple of 10.

The algorithm works for many other cards including an 18-digit Tesco card:

<u>6</u> 3 <u>4</u> 0 <u>0</u> 4 <u>0</u> 2 <u>4</u> 0 <u>8</u> 4 <u>0</u> 7 <u>1</u> 0 <u>5</u> 1

As before you have:

$$(1 + 2) + 8 + 0 + 0 + 8 + (1 + 6) + 0 + 2 + (1 + 0) + 3 + 0 + 4 + 2 + 0 + 4 + 7 + 0 + 1 = 50$$

Leonhard Euler

16 provides another excuse to tell you about Leonhard Euler, the eighteenth-century Swiss mathematician and number theorist. Euler was one of the world's greatest mathematicians. He invented several mathematical symbols that are now standard including the one for π (or, to be precise, it was Euler's use of the symbol that established it as standard notation, since there were a few earlier uses of the same symbol). The ratio of the circumference of a circle to its diameter, $\pi = 3.14159 \ldots$ the digits go on in a random fashion forever and they never repeat. A computer, which must be as big as a garden shed, has calculated the decimal places to 13 trillion places – yes, that's

13,000,000,000,000 – and no one digit appears significantly more than any other. Imagine how thrilled Euler would have been with such a computer.

Euler made so many discoveries about the transcendental number *e* that it is known as Euler's number. Like π, the decimal places of *e* go on forever but *e* = 2.718 is a good enough approximation for most calculations. Most people know $^{22}/_7$, the simple fraction for π, but $^{878}/_{323}$ is a memorable fraction for *e*. The top and bottom numbers are palindromic and the difference between them is 555.

Euler, like almost every mathematician, was also fascinated by prime numbers and his formula for generating these addictive numbers will appear in a later chapter.

Euler invented the symbol *i* for the imaginary number $\sqrt{-1}$, which is an impossible concept for most ordinary mortals but is hugely important in aerodynamics and other branches of engineering.

And there *is* one very tiny scrap of this great man's knowledge that we can all understand. He proved that the only solution to the equation $a^b = b^a$ is $2^4 = 4^2 = 16$. Of course, perhaps we couldn't prove it was the *only* solution.

Some Interesting Temperatures

When it is 16°C (61°F) in the author's home town of Montrose, Scotland, people greet each other in the High Street with 'Lovely day! But there's still that cool breeze off the sea!' In London at 28°C (82°F) workers head for the nearest park to eat their lunchtime sandwiches in the sun.

The village of Oymyakon in Siberia is said to be the coldest populated place in the world. In 1924, there was a record low of –71.2°C (–96.2°F). However, the average winter temperature is –40°. Is that Centigrade or Fahrenheit? It doesn't matter. At that temperature only, –40°F and –40°C are exactly the same.

Leyland Numbers

Leyland numbers are the newest numbers in this book as they were invented only this century by British number theorist Paul Leyland.

When Euler in the previous chapter proved that the *only* solution to $a^b = b^a = 16$, did he not experiment with $x^y + y^x$ and see where it got him? Out of this neat little formula, Paul Leyland has invented a new set of numbers.

But neither x nor y can be 1. Think about it: $1^4 + 4^1 = 5$, $7^1 + 1^7 = 8$ or $a^1 + 1^a = a + 1$, and you would produce all the numbers from here to infinity. Here is a list of the first 20 Leyland numbers and how they are calculated:

x	y	$x^y + y^x$
2	2	8
2	3	17
2	4	32
3	3	54
2	5	57
2	6	100
3	4	145
2	7	177

x	y	$x^y + y^x$
2	8	320
3	5	368
4	4	512
2	9	593
3	6	945
2	10	1124
4	5	1649
2	11	2169
3	7	2530
2	12	4240
4	6	5392
5	5	6250

And so they go on forever.

In the above list, only 17 and 593 are prime but already the enormous $2638^{4405} + 4405^{2638}$ has been proved by computer methods to be a very large prime number.

17

17 is the only prime number of the form $p^q + q^p$ where p and q are also prime numbers, namely 2 and 3. In other words, $17 = 2^3 + 3^2$.

$17^3 = 4913$ and $4 + 9 + 1 + 3 = 17$

The reciprocal of 17 is $^1/_{17} = 0.0588235294117647$ ... and repeats after 16 digits. Later in the book you will find that the reciprocals of some prime numbers repeat after a short run of digits. Others are very long. But they are never longer than one digit fewer than the number itself. So the reciprocal of 17 is at its maximum length with 16 digits. Earlier, the reciprocal of 7 also had a maximum length reciprocal with 6 digits.

19

The prime number 19, when multiplied, makes such a lovely pattern.

19 × 1 = 19 and 1 + 9 = 10 and 1 + 0 = 1
19 × 2 = 38 and 3 + 8 = 11 and 1 + 1 = 2
19 × 3 = 57 and 5 + 7 = 12 and 1 + 2 = 3
19 × 4 = 76 and 7 + 6 = 13 and 1 + 3 = 4
19 × 5 = 95 and 9 + 5 = 14 and 1 + 4 = 5
19 × 6 = 114 and 11 + 4 = 15 and 1 + 5 = 6
19 × 7 = 133 and 13 + 3 = 16 and 1 + 6 = 7
19 × 8 = 152 and 15 + 2 = 17 and 1 + 7 = 8
19 × 9 = 171 and 17 + 1 = 18 and 1 + 8 = 9
19 × 10 = 190 and 19 + 0 = 19 and 1 + 9 = 10

There is not a great deal more to say about 19 itself but, wow, it has a great reciprocal. After 7 and 17, it is the 3rd prime number whose reciprocal is maximum length: 1 over 19 goes the full 18 decimal places before it repeats.

The Reciprocal of 19

1 ÷ 19 = 0.052631578947368421 …

You can multiply 052631578947368421 by any number from 2 to 18 by finding the correct starting place, reading off the digits until you get to the end of the number, circling back to the beginning and stopping where you first started.

For example, if you want to multiply by 5, choose the first 5 with the 2 after it. If you want to multiply by 15 choose the second 5 with the larger digit 7 after it. Likewise, for 8 and 18, choose the 8 that is followed by 4 for multiplying by 8; choose the 8 that is followed by 9 for multiplying by 18.

0526315 78947368421 × 3 = 157894736842105263
052631578947368421 × 13 = 684210526315789473

Try multiplying the number by 19 and you get a row of 9s. Split the 18 digits in half and add the two 9-digit numbers. You will not be too surprised at the result.

Divisibility by 19

Finding out if a number divides by 19 is surprisingly easy. Double the last digit, add it to the rest of the number. Repeat this until you can go no further. Here are some examples:

1. Does 9172 divide by 19?

    917 + 4 = 921
    92 + 2 = 94
    9 + 8 = 17

    So 9172 is not divisible by 19

2. Is 2888 divisible by 19?

   288 + 16 = 304
   30 + 8 = 38

   38 divides by 19 so 2888 also divides by 19.

3. Is 1,000,008 divisible by 19?

   100,000 + 16 = 100,016
   10,001 + 12 = 10,013
   1001 + 6 = 1007
   100 + 14 = 114
   11 + 8 = 19

   Hey presto! 1,000,008 is divisible by 19.

27

You should have been informed earlier. Sorry about that! In the section on the Collatz conjecture ('A Twentieth-century Conjecture'), you were invited to choose a number, divide by 2 if it was even, or treble it and add 1 if it was odd. You had to keep doing this until, eventually, you ended up with 1. Well, if you chose 27, perhaps you have been tearing your hair out. You see, 27 needs 111 steps to reach 1 and many of the numbers along the way are very big. So only choose 27 if you are a glutton for punishment – and if, and only if, you really have nothing else to do.

Otherwise 27 is quite a nice number. $27 = 3^3$ and $27^3 = 19,683$ and the digits of 19,683 add up to 27.

The reciprocal of 27 (1 over 27) is a never-ending decimal 0.037037037 … and guess what the reciprocal of 37 is? Yes, another little bit of magic – it's 0.027027027 …

Take any 3-digit multiple of 27, say 27 × 8 which is 216, turn the digits around in a clockwise direction to 162 or 621 and you will find that they also divide by 27. And what other number has the same property? Yes, you guessed it! It's 37.

28

28 deserves our respect because it is triangular, hexagonal and perfect.

There are 28 dominoes in a box. Letting B stand for blank, you have:

B-B
1-1   1-B
2-2   2-1   2-B
3-3   3-2   3-1   3-B
4-4   4-3   4-2   4-1   4-B
5-5   5-4   5-3   5-2   5-1   5-B
6-6   6-5   6-4   6-3   6-2   6-1   6-B

This is $1 + 2 + 3 + 4 + 5 + 6 + 7 = 28$. Or, doing it the Gauss way, $7 \times 8 \div 2 = 28$.

Domino Fractions

Remove all the doubles and blanks from the box and you are left with 15 dominoes, 1-2, 1-3, 1-4, 1-5, 1-6, 2-3, 2-4, 2-5, 2-6, 3-4, 3-5, 3-6, 4-5, 4-6 and 5-6.

Substituting digits for spots, regard each domino as a fraction.

$$\frac{3}{4} \quad \frac{1}{4} \quad \frac{3}{6} \quad \frac{1}{2} \quad \frac{2}{4}$$

$$\frac{5}{6} \quad \frac{2}{6} \quad \frac{1}{3} \quad \frac{4}{5} \quad \frac{1}{5}$$

$$\frac{4}{6} \quad \frac{1}{6} \quad \frac{2}{3} \quad \frac{3}{5} \quad \frac{2}{5}$$

Each row of domino fractions adds to 2½.

29

$2n^2 + 29$ generates prime numbers for all values of n from 0 to 28. They start 29, 31, 37, 47, 61, 79, 101, 127 ... with the last ones being 1487 and 1597.

30 and the Giuga Numbers

Many people, especially schoolchildren, think that maths is an old-fashioned subject in which things were discovered and proved a long time ago, and everything in maths is 'done and dusted'. Here is a set of rare numbers, where only 11 are known so far.

In the middle of the twentieth century Giuseppe Giuga, an Italian mathematician, identified this new set of numbers. He wrote down the prime factors of 30, which are 2, 3 and 5. He added their reciprocals, ½ + ⅓ + ⅕, getting ³¹⁄₃₀ and then he multiplied the reciprocals ½ × ⅓ × ⅕, getting ¹⁄₃₀ and subtracted this from the first result, getting 1. (This also has the consequence that when you divide 30 by each of its factors, 2, 3 and 5, then subtract 1, you get a multiple of the factor. So: $30/2 - 1 = 7 \times 2$; $30/3 - 1 = 3 \times 3$; $30/5 - 1 = 1 \times 5$.)

These were simple calculations but Giuga discovered that no other numbers, given this treatment, worked again until 858. The prime factors of 858 are 2, 3, 11, 13 and:

$$(½ + ⅓ + ¹⁄₁₁ + ¹⁄₁₃) - (¹/_2 \times ¹/_3 \times ¹⁄₁₁ \times ¹⁄₁₃) = 1$$

The next one to work was 1722 with its prime factors 2, 3, 7 and 41, and then far along the number line, 66,198 with prime

·factors 2, 3, 11, 17 and 59 gave the satisfying result of 1 again. A fifth one with 10 digits and 6 factors was found, and then 6 others of enormous size.

It is not known if this set of numbers is infinite or whether there are any odd Giuga numbers. So the research is ongoing. Could it be that Giuga numbers are not forever?

31

$$31 = 5^0 + 5^1 + 5^2$$
$$31 = 2^0 + 2^1 + 2^2 + 2^3 + 2^4$$

The reciprocal of 31 is half the maximum length with 15 digits:

$\frac{1}{31} = 0.032258064516129 \ldots$

Split it into 3 bits and add:

03225
80645
16129

There's the 99999 string again.

31 and Its Big Brothers

31 is prime and so are 331, 3331, 33331, 333331, 3333331 and 33333331. But the prime pattern stops there because 333333331 divides by 17.

37

Because 37 is related to the repunit 111 (37 = 111 ÷ 3) it sometimes has repetitive ways:

$37 \times 3 = 111$
$37 \times 6 = 222$
$37 \times 9 = 333$

Keep going up the 3 times table to:

$37 \times 27 = 999$

And continue going up in 3s to see another pattern emerge:

$37 \times 30 = 1110$
$37 \times 33 = 1221$
$37 \times 36 = 1332$
$37 \times 39 = 1443$
$37 \times 42 = 1554$

And so on.

37 and the Centred Hexagonal Numbers

This is as good a place as any to discuss the centred hexagonal numbers. Each individual hexagon has 6, 12, 18, 24, 30, etc. spots, with 1, of course, the centre of attention in the middle. The above spotty diagram shows 37, the fourth one, but you can see the first, second and third inside. Counting the spots, they are 1, 1 + 6, 1 + 6 + 12 and 1 + 6 + 12 + 18, giving 1, 7, 19 and 37.

The formula for centred hexagonal numbers is:

$$3n^2 - 3n + 1$$

Let's find the 5th one.

When n = 5, $3n^2 - 3n + 1 = 3 \times 25 - 15 + 1 = 61$. So the list goes on forever, 1, 7, 19, 37, 61, 91, 127, 169, 217, 271 …

Look at this:

$$1 + 7 = 8 = 2^3$$
$$1 + 7 + 19 = 27 = 3^3$$
$$1 + 7 + 19 + 37 = 64 = 4^3$$
$$1 + 7 + 19 + 37 + 61 = 125 = 5^3$$

And further on:

$$1 + 7 + 19 + 37 + 61 + 91 + 127 + 169 + 217 + 271 = 1000$$
$$= 10^3$$

The sum of the first n centred hexagonal numbers is n^3.

Magic Hexagons

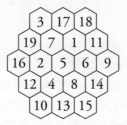

This image is of a hexagon of order 3. Using the same formula as above, $3n^2 - 3n + 1$, we see that we need from 1 to $(3 \times 3^2 - 3 \times 3 + 1) = 19$ numbers to fill it.

Go to any corner number, for example 18, and add in any direction, $18 + 11 + 9$ or $18 + 1 + 5 + 4 + 10$ or $18 + 17 + 3$. In each case you get 38. Likewise, for any other number on the edge, you get 38, for example, $12 + 4 + 8 + 14$, $12 + 2 + 7 + 17$ and $12 + 16 + 10$.

41

There are rich pickings in Euler's famous formula for generating prime numbers, $x^2 + x + 41$. Every number from $x = 0$ to $x = 39$ substituted in the formula yields a prime number, giving the long string 41, 43, 47, 53, 61, 71, 83, 97, 113, 131, up to 1601, but it fails at $x = 40$ because $40^2 + 40 + 41 = 1681$, which is 41^2. It also fails, of course, for $n = 41$ and every multiple of 41.

But the formula is by no means finished.

When $n = 42$, $n^2 + n + 41 = 1764 + 42 + 41 = 1847$ (prime)
When $n = 43$, $n^2 + n + 41 = 1849 + 43 + 41 = 1933$ (prime)
When $n = 44$, $n^2 + n + 41 = 1936 + 44 + 41 = 2021$ (47×43)
When $n = 45$, $n^2 + n + 41 = 2025 + 45 + 41 = 2111$ (prime)
When $n = 46$, $n^2 + n + 41 = 2116 + 46 + 41 = 2203$ (prime)

And so on.

Later, when $n = 1000$, the formula yields 1,001,041, which is prime and when $n = 2000$, you get 4,002,041, also prime. But when $n = 3000$, the result is not prime.

The formula continues to be a fertile source of prime numbers on its journey along the number line.

In 1772, Euler proved that the 10-digit Mersenne number $2^{31} - 1 = 2,147,483,647$ was prime.

5-Digit Multiples of 41

If any multiple of 41 has 5 digits, for example 24,067 (41 × 587), when it is rotated in a clockwise direction, the new numbers will also be divisible by 41:

$40,672 \div 41 = 992$
$06,724 \div 41 = 164$
$67,240 \div 41 = 1640$
$72,406 \div 41 = 1766$
$24,067 \div 41 = 587$

Start with 41

41 is the start of a prime number chain that is 39 numbers long:

$41 + 2 = 43$
$43 + 4 = 47$
$47 + 6 = 53$
$53 + 8 = 61$
$61 + 10 = 71$
$71 + 12 = 83$
$83 + 14 = 97$
$97 + 16 = 113$

And so on until lastly:

1523 + 78 = 1601

But:

1601 + 80 = 1681

This is not prime because 1681 = 41^2.

The Reciprocal of 41

$1/41$ = 0.0243902439 … It repeats after only 5 decimal places. (Note the total of the 5 digits.)

47, 497, 4997 *et al.*

47 + 2 = 49 and 47 × 2 = 94
497 + 2 = 499 and 497 × 2 = 994
4997 + 2 = 4999 and 4997 × 2 = 9994

Keep packing 9s into the centre of 47 forever and the pattern will never fail.

The Magic 48^2

48^2 = (drum roll required)

$2^2 + 5^2 + 8^2 + 11^2 + 14^2 + 17^2 + 20^2 + 23^2 + 26^2$

(Notice the 2, 5, 8, 11, 14 ... sequence.) Just sit back and admire before you check it.

70 and Other
Weird Numbers

These numbers don't deserve their name. They are rare rather than bizarre.

All weird numbers are abundant (see 'Abundant Numbers') but in spite of the fact that 12, 18, 20, 24, 30, 36, 40, 42, 48, 54, 56, 60 and 66 all appear on the abundant list before 70, 70 is the first weird number. So why is 70 a weird, abundant number?

The divisors of 24, for example, are 1, 2, 3, 4, 6, 8 and 12, making the abundance of 24 equal to 36, but within that group of divisors is a subset, 4, 8 and 12, which adds to *exactly* 24.

Take another example, like 60. The divisors of 60 are 1, 2, 3, 4, 5, 6, 10, 12, 15, 20, and 30, giving an abundance of 108, but again there is a subset within that group of 30, 20 and 10, which adds to 60.

This happens with almost all of the abundant numbers and 70 is the first exception. The divisors of 70 are 1, 2, 7, 10, 14, and 35, adding to 74, and there is no subset within the group that adds to 70. So that is why 70 is weird.

The weird numbers under 10,000 are an exclusive little group. They are 70, 836, 4030, 5830, 7192 and 9272. There are only 6 of them and they are all even.

It is known that weird numbers go on forever but it is not known if there are any odd weirdies. If so, they must be larger than 4 trillion.

71

There are some interesting little snippets about numbers in the 70s. Here are a couple for 71:

$71^2 = 7! + 1$ and $71^3 = 357,911$

357,911 is made up of the odd numbers from 3 to 11.

74

$74^2 = 5476$ and $74^4 = 29,986,576$

And guess what these 12 digits add up to?

75

75 is a little gold mine of prime numbers.

$$75 - 2^1 = 73$$
$$75 - 2^2 = 71$$
$$75 - 2^3 = 67$$
$$75 - 2^4 = 59$$
$$75 - 2^5 = 43$$
$$75 - 2^6 = 11$$

81

81 has an interesting reciprocal.

$^1/_{81}$ = 0.0123456790123456790123 …

Note that the 8 is missing.

Take one period of the repeating reciprocal, 012345679, and multiply it by this selection of numbers starting at 19 and with a gap of 9:

```
012345679 × 19 = 234567901
012345679 × 28 = 345679012
012345679 × 37 = 456790123
012345679 × 46 = 567901234
012345679 × 55 = 679012345
012345679 × 64 = 790123456
012345679 × 73 = 901234567
```

In every product, 8 has gone on holiday.

Now take 12345679 again and multiply it with numbers, this time with a gap of 27. You may miss out zero this time as it is not part of the pattern.

```
12345679 × 3 = 037037037
12345679 × 30 = 370370370
12345679 × 57 = 703703703
```

12345679 × 6 = 074074074
12345679 × 33 = 407407407
12345679 × 60 = 740740740

12345679 × 12 = 148148148
12345679 × 39 = 481481481
12345679 × 66 = 814814814

12345679 × 15 = 185185185
12345679 × 42 = 518518518
12345679 × 69 = 851851851

12345679 × 21 = 259259259
12345679 × 48 = 592592592
12345679 × 75 = 925925925

12345679 × 24 = 296296296
12345679 × 51 = 629629629
12345679 × 78 = 962962962

The above 6 groups contain patterns:

1. The middle multiplier in each of the above 6 groups equals the sum of the digits of each result in that group.

    For example, 48 = 2 + 5 + 9 + 2 + 5 + 9 + 2 + 5 + 9.

2. The sum of the multipliers of each group equals the sum of all the digits of the 3 results in that group.

12345679 reacts differently when multiplied by 9, 18, 27 and, indeed, any multiple of 9. First of all, you get strings of the same number, for example, 12345679 × 9 = 111111111 and 12345679 × 81 = 999999999. But after that the identical strings are still there for multiples of 9 but the result is broken up a bit. For example, multiplying by 135 gives 1666666665, but add 1 and 5 and you have a string of 6s again.

89

89 and the next prime after that, 97, are the first pair of consecutive primes that differ by 8.

The reciprocal $1/89$ starts 0.011235955056179 ... and it repeats after 44 decimal places. This reciprocal looks exciting because the first 6 digits, 0, 1, 1, 2, 3, 5 are the first numbers in the Fibonacci sequence. But, alas, it doesn't look as if there is any magic stuff this time because, as you know, the famous sequence goes 0, 1, 1, 2, 3, 5, 8, 13, 21, 34, 55, 89, 144 ...

However, here are the Fibonacci numbers written in a decimal fashion:

0.0
0.01
0.001
0.0002
0.00003
0.000005
0.0000008
0.00000013
0.000000021
0.0000000034
0.00000000055
0.000000000089

Now add them together and you get 0.011235955039, which is rather similar to the above reciprocal. If you continue to attach more and more rows starting with 0.0000000000144, the addition will get even more like the 89 reciprocal. This strange interrelation between Fibonacci and 89 is number magic at its most spectacular.

89 and Others Going Up in Power

$89 = 8^1 + 9^2$
$135 = 1^1 + 3^2 + 5^3$
$175 = 1^1 + 7^2 + 5^3$
$598 = 5^1 + 9^2 + 8^3$

97

The reciprocal of 97 starts 0.010309278 ... and goes on for its maximum length. The party piece of some mathematicians is to rattle off all 96 of them by heart. You might wonder if mathematicians get invited to many parties.

Reciprocals

We've already looked at a few interesting reciprocals – here is an overview of how they work.

The decimal reciprocal of numbers whose prime factors are *only* 2 and/or 5 can be worked out exactly. For example the reciprocal of 25 is $^1/_{25}$ = 0.04 and the reciprocal of 128 is $^1/_{128}$ = 0.0078125. But other reciprocals can't be exactly expressed in a finite number of decimal places. So apart from 2 and 5, the reciprocals of all prime numbers go on forever but repeat after a certain number of digits. The repeated part of the recurring decimal is called the 'period' or the 'repetend'. This repetend sometimes has the maximum number of digits that is one fewer digit than the prime number itself or, sometimes, only a fraction of this maximum number.

Take the reciprocal for 7. The repetend is 142857. Add the digits. Do the same for 13, whose repetend is 076923, and 43, with its repetend of 023255813953488372093. In each case *the digits add to a multiple of 9*. Try any other one. Note the last column in the following table.

The table does not include the digits of each prime reciprocal but quite a few have been calculated for you on previous pages. Write down the repetend, split it in half and add.

For example, $^1/_{17}$ = 0.0588235294117647 … Here you have 05882352 + 94117647 = 99999999.

$1/73$ has a short repetend, 0.01369863. Split it in half to get 0136 + 9863 = 9999.

But if the repetend has an odd number of digits, say 15 as in $1/31$ = 0.032258064516129, the digits have to be split into thirds: 03225 + 80645 + 16129 and you will get what you expected.

Try it for several others.

The Reciprocals of the Prime Numbers from 7 to 97

Prime Number (P)	Formula for repetend	Number of dec pl	Sum of digits
7	P – 1	6	27
11	$1/5$ (p – 1)	2	9
13	½ (p – 1)	6	27
17	p – 1	16	72
19	p – 1	18	81
23	p – 1	22	99
29	p – 1	28	126
31	½ (p – 1)	15	54
37	$1/12$ (p – 1)	3	9
41	⅛ (p – 1)	5	18

Prime Number (P)	Formula for repetend	Number of dec pl	Sum of digits
43	½ (p − 1)	21	90
47	p − 1	46	207
53	¼ (p − 1)	13	63
59	p − 1	58	261
61	p − 1	60	270
67	p − 1	66	288
71	½ (p − 1)	35	126
73	⅑ (p − 1)	8	36
79	⅙ (p − 1)	13	54
83	½ (p − 1)	41	171
89	½ (p − 1)	44	198
97	p − 1	96	423

Just a Thought

In 1588, Pietro Cataldi proved that $2^{19} − 1 = 524,287$ is a prime number with no calculating aid whatsoever and, in 1772, Euler used only clever reasoning and trial division to prove that $2^{31} − 1 = 2,147,483,647$ is prime. And, of course, there are all these supermegaprimes being discovered in the twenty-first century by computer experts on a regular basis.

Have they, do you think, also calculated the reciprocals of these new primes? Have they added the digits of the repetends and have they done the split business and been rewarded with mega printouts of 9s, 9s and more 9s? We may never know, but it is fun to imagine.

100

There is a range of methods to make 100 using the digits from 1 to 9:

123 + 4 − 5 + 67 − 89 = 100
123 − 4 − 5 − 6 − 7 + 8 − 9 = 100
1 + 2 + 34 − 5 + 67 − 8 + 9 = 100
1 + 23 − 4 + 56 + 7 + 8 + 9 = 100
1 + 2 + 3 − 4 + 5 + 6 + 78 + 9 = 100
123 − 45 − 67 + 89 = 100
98 − 76 + 54 + 3 + 21 = 100

And there are many others if you allow multiplication and division.

101 and Its Other Palindromic Pals

One Hundred and One Dalmatians featured one of Disney's most notorious villains, Cruella De Vil, who kidnapped 101 sweet little spotted puppies to make them into fur coats.

As already mentioned, a palindrome is any word, phrase or number that reads the same backwards as forwards, so 101 is called a palindromic prime. It is only the second smallest one as little 'legs 11' has the honour of being the only 2-digit palindromic prime. 101 isn't very interesting except that a number with digits ab multiplied by 101 is abab. For example, $46 \times 101 = 4646$.

But 101 does have a great-sounding reciprocal: zero zero nine nine, zero zero nine nine, zero zero nine nine …

Its palindromic bigger brother, 10101, is not prime. (A glance at it tells you immediately that it divides by 3.) In fact, 10101 is full of factors and equals $3 \times 7 \times 13 \times 37$ so you can have some fun with it. If you look back to the chapter on '7' you will see why $49 \times 37 \times 13 \times 3 \times 7$ is 494949. In the same way as 101, ab $\times$ 10101 is ababab.

105

Do exactly the same calculations as for 75 above and from $105 - 2^n$ you will get some more primes: 103, 101, 97, 89, 73 and 41. $45 - 2^n$ yields 43, 41, 37, 29 and 13. From $21 - 2^n$, you get 19, 17, 13 and 5 and from $15 - 2^n$, 13, 11 and 7.

109

Here's a curiosity involving the number 109, discovered by Rick Toews at the beginning of the twenty-first century. The reciprocal of 109 goes the full 108 digits and they have been written out so that, once again, you can see 2 halves adding to 999999999 … (see the 'Reciprocals' chapter).

0.00917431192660550458715596330275229357798165137614678899082568807339449541284403669724770642201834862385321

This time there is something else that is very interesting. You can see the beginning of the Fibonacci sequence in the last 6 digits appearing in *reverse* order starting from the *end* of the decimal (i.e. 0, 1, 1, 2, 3, 5, 8 appears as 853211). After that, the digits do not appear to be anything like the Fibonacci numbers, 13, 21, 34, 55, 89, 144, etc. However, do this:

At 13, write down the 3 and carry the 1, so 21 is now 22.
Write down 2 and carry 2 so 34 is now 36.
Write down 6 and carry 3 so 55 is now 58.
Write down 8 and carry 5 so 89 is now 94.
Write down the 4 and carry 9 so 144 is now 153.
Write down the 3 and carry 15 so 233 is now 248.
Write down the 8 …

Writing out the digits again in reverse order and attaching the 'write down' digits you get … 8348623853211 and you can keep going as long as you like.

Compare these 13 digits with the last 13 digits of 109's reciprocal. They are identical.

Once again, you are witnessing a numerical phenomenon.

112 and the
Rest of the Family

$112 \rightarrow (1 + 1 + 2)^2 = 16$ and $112 \div 16 = 7$
$162 \rightarrow (1 + 6 + 2)^2 = 81$ and $162 \div 81 = 2$
$243 \rightarrow (2 + 4 + 3)^2 = 81$ and $243 \div 81 = 3$
$324 \rightarrow (3 + 2 + 4)^2 = 81$ and $324 \div 81 = 4$

The others in the family, 392, 405, 512, 605, 648, 810 and 972, behave in the same way.

118

118 = 15 + 40 + 63
118 = 14 + 50 + 54
118 = 18 + 30 + 70
118 = 21 + 25 + 72

So what? Well, $15 \times 40 \times 63 = 14 \times 50 \times 54 = 18 \times 30 \times 70 = 21 \times 25 \times 72 = 37{,}800$.

The 123 Curiosity

Write down any number. Make it as long as you like as there is very little calculation to be done. For example:

4,583,276,143,999

Write down the number of even digits, odd digits and the total number of digits.

5 | 8 | 13

Write this down as a new number and repeat the above instructions.

5813
1 | 3 | 4
134
1 | 2 | 3
123

Whatever number you start with, you'll always end with 123. (This works for vowels/consonants in a sentence, boys/girls in a class, etc. Otherwise, you can simply start at the second line and write down any 2 numbers and the total; proceed as before and the result is always 123.)

127: De Polignac's Great Disappointment

Alphonse de Polignac, a nineteenth-century French aristocrat and mathematician, thought he had discovered something new and exciting involving prime numbers.

He thought that any odd number, except 1, could be expressed as a power of 2 plus a prime number.

$3 = 2^0 + 2$
$5 = 2^1 + 3$
$7 = 2^2 + 3$
$9 = 2^2 + 5$
$11 = 2^2 + 7$
$13 = 2^3 + 5$

And on he went, confidently racing through to:

$99 = 2^5 + 67$
$101 = 2^6 + 37$

And all the odd numbers up to 127. He got stuck with this one but kept going.

$129 = 2^5 + 97$
$131 = 2^6 + 67$

Eventually he reached 999 and that worked too.

$$999 = 2^5 + 967$$

(He double-checked that 967 was prime.)

He went on and on – there was no stopping him – up to odd numbers like 8209.

$$8209 = 2^{13} + 17$$

He kept going but eventually, he returned to 127 to see where he must have made a silly error. And do you know what? *127 does not work.*

He tried it again.

$$127 = 2^0 + 126$$
$$127 = 2^1 + 125$$
$$127 = 2^2 + 123$$
$$127 = 2^3 + 119 \text{ (119 divides by 7)}$$
$$127 = 2^4 + 111$$
$$127 = 2^5 + 95$$
$$127 = 2^6 + 63$$

And that's it! We can't go any further because $2^7 = 128$.

Alphonse was very disappointed. This time, the beautiful numbers had let him down and his conjecture had failed.

Up in number heaven, de Polignac will be delighted to know that the other conjecture for which he is remembered has never been disproved, although it has never been proved either. This one is about prime numbers again and states that there are infinitely many gaps between prime numbers of size n, where n is an even number. If de Polignac's conjecture is true, it would cover twin primes (n = 2) as well as cousin primes (n = 4), sexy primes (n = 6) and every other even value of n.

132

The 2-digit numbers you can make from 132 are 13, 12, 32, 31, 23 and 21. Guess what you get when you add them?

136

A little bit of tit for tat. $1^3 + 3^3 + 6^3$ doesn't make 136 as you might have expected. It adds to 244, but $2^3 + 4^3 + 4^3 = 136$.

144

$144 = 12^2$ and is the 12th entry and the only square number in the Fibonacci sequence.

Factorial Fun with 145

$145 = 1! + 4! + 5!$

153 and the
Narcissistic Numbers

According to Greek mythology, Narcissus was a beautiful youth who saw his reflection in a pool of water and fell in love with his own image. He never left the pool and wasted away. The narcissus flower sprang up where he died.

It could be said that narcissistic numbers are obsessed with themselves. For example, 153 has 3 digits: 1, 5 and 3. It so happens that $1^3 + 5^3 + 3^3 = 1 + 125 + 27 = 153$.

So a narcissistic number is an n-digit number equal to the sum of its digits raised to the nth power, that is, the power corresponds to the number of digits.

$$1634 = 1^4 + 6^4 + 3^4 + 4^4 = 1 + 1296 + 81 + 256 = 1634$$
$$1,741,725 = 1^7 + 7^7 + 4^7 + 1^7 + 7^7 + 2^7 + 5^7$$
$$= 1 + 823,543 + 16,384 + 1 + 823,543 + 128 + 78,125$$
$$= 1,741,725$$

Excluding the trivial 1-digit numbers, there are only 88 of these narcissistic numbers and the biggest 2 have a whopping 39 digits. At this stage, you must believe that whole numbers are truly magical and maybe the Pythagoreans were right after all. OK, here is one of the 39s: 115,132,219,018,763,992,565,095,597,973,971,522,400 – but without a supercomputer, it would be impossible to check.

However, the fun is having a fairly challenging one to check by yourself.

How about 54,748?

The Narcissistic Numbers are also called the Armstrong Numbers or Pluperfect Digital Invariants (PPDIs, not to be confused with PDDIs, more about which in the '3435 and Münchhausen' section.)

The Friedman Numbers

The Friedman numbers are named after Erich Friedman, Associate Professor of Mathematics at Stetson University, Florida, and ardent maths puzzle expert.

A Friedman number is as self-obsessed as the narcissistic numbers in the previous chapter because it is expressed using all of its own digits and any or all of the mathematical operations, +, −, ×, ÷, powers, brackets and concatenations. Concatenations? An example of this is: $13{,}243 = 41 \times 323$ where the separate digits are linked or gathered together to form new numbers. Likewise $126 = 6 \times 21$ and $3159 = 351 \times 9$ have been concatenated.

In the first 10,000 numbers there are hundreds of Friedman numbers, all worked out by Erich and his colleagues. Other simple examples of these clever numbers are $2509 = 50^2 + 9$, $127 = 2^7 - 1$ and $39328 = 2 \times 3^9 - 38$.

However, Erich's colleague, Mike Reid, suggested that it would be 'nicer' if the digits were in the correct order and so began Nice Friedmans. The nice ones are not quite so numerous as the 'not so nice' Friedmans but they are more spectacular. Sit back and enjoy a few of the infinite number of Nice Friedmans (before you check them):

$2737 = (2 \times 7)^3 - 7$
$35721 = 3^5 \times 7 \times 21$
$216003 = 2 + 1 + (60 + 0)^3$
$46630 = 4 + 6^6 - 30$
$117652 = 1 - 1 + 7^6 + 5 - 2$

The following pandigits are not nice but wow, aren't they clever!

$123456789 =$
$\{(86 + 2 \times 7)^5 - 91\} \div 3^4$
and
$987654321 =$
$\{8 \times (97 + 6 \div 2)^5 + 1\} \div 3^4$

These are quite easy to check. Remember to do × and ÷ before + and −.

And who would have thought that a repunit would be a Friedman number! The 11-legged repunit $11111111111 = \{(11 - 1)^{11} - 1 \times 1\} \div (11 - 1 - 1)$.

Another of Erich's maths friends, Michael Brand, claims that, as $n \to \infty$, the density of Friedman numbers approaches 1. This is maths speak for his conjecture that almost every enormous number is a Friedman number. So if you *were* to write down a random 100-digit number, he is claiming that the probability of it being a Friedman number is very high. However, the probability of you being able to prove that your chosen number is a Friedman is very low indeed!

Of course, there will always be some numbers, such as all powers of 10, like 1000, 1 million or a googol (1 with 100 zeros), that are not Friedman numbers.

197

Take 197 and add the digits:

$1 + 9 + 7 = 17$

Now keep going in Fibonacci fashion:

$9 + 7 + 17 = 33$
$7 + 17 + 33 = 57$
$17 + 33 + 57 = 107$
$33 + 57 + 107 = 197$

199

199 is prime and so are 919 and 991.

220 and Amicable Pairs

220 has the honour of being the shorter friend in the very first pair of amicable numbers. 220 and 284 are best buddies because the sum of the divisors of 220 (including 1 but not the number itself) is 284, and the sum of the divisors of 284 is 220. Get that? Perhaps not. It is easier to see in the table.

Divisors of 220	Divisors of 284
1	1
2	2
4	4
5	71
10	142
11	
20	
22	
44	
55	
110	
Total 284	Total 220

Pure magic, you must agree. Over 2 millennia ago the Pythagoreans knew about this pair and considered the numbers to be the symbols of friendship.

The next loving pair are 1184 and 1210. They were discovered by a 16-year-old Italian schoolboy about 160 years ago, long after the famous mathematicians had missed this pair and gone on to multi-digit pairs much further along the number line.

Divisors of 1184	Divisors of 1210
1	1
2	2
4	5
8	10
16	11
32	22
37	55
74	110
148	121
296	242
592	605
Total 1210	Total 1184

Just as mathematicians have searched long and hard for more and more prime numbers and perfect numbers, the same has been done for amicable pairs. Until the advent of computers, the amicable numbers were a select little group but now the list is in the hundreds and growing.

There are only thirteen pairs that are under 100,000. They are listed below.

A	B
220	284
1184	1210
2620	2924
5020	5564
6232	6368
10,744	10,856
12,285	14,595
17,296	18,416
63,020	76,084
66,928	66,992
67,095	71,145
69,615	87,633
79,750	88,730

And on and on they go with the number of digits getting more and more mind-blowing, especially when you have to find all their divisors.

There is no known formula that will create *all* the pairs but one formula, discovered by Thabit ibn Qurra, an Arab mathematician in the ninth century, creates 2 of the smaller pairs. In the sixteenth century, another Arab mathematician, Ibn al-Banna, recorded the very impressive pair of amicable numbers 17,296 and 18,416, which is the eighth pair in the official table. In the seventeenth century, Pierre de Fermat rediscovered Thabit's formula and took it further, and Leonard Euler, the eighteenth-century Swiss mathematician, used other methods to add several more amicable pairs to the list.

Here is Thabit's simple formula, which he used to create a few of the more manageable friendly pairs.

Let's make 3 numbers, a, b and c, from the following 3 formulae.

$$a = 3 \times 2^n - 1$$
$$b = 3 \times 2^{n-1} - 1$$
$$c = 9 \times 2^{2n-1} - 1$$

Now, first of all, we have to choose a value for n so that a, b and c will be prime numbers. This is guesswork. We don't know which values of n will work and which won't work, so we might as well choose the easiest value, n = 2.

When n = 2, $a = 3 \times 2^2 - 1 = 3 \times 4 - 1 = 11$
When n = 2, $b = 3 \times 2^{2-1} - 1 = 3 \times 2^1 - 1 = 5$
When n = 2, $c = 9 \times 2^{2 \times 2 - 1} - 1 = 9 \times 2^{4-1} - 1$
$= 9 \times 2^3 - 1 = 9 \times 8 - 1 = 71$

Hurrah! a, b and c are 11, 5 and 71 and are all prime numbers!

But there's more to come in order to create your pair of amicable numbers.

Taking 11, 5 and 71 for a, b and c, and n = 2 as before, use the 2 formulae $2^n \times a \times b$ and $2^n \times c$ and your amicable pair will appear.

$2^2 \times 11 \times 5 = 220$ and $2^2 \times 71 = 284$

If, again, you let n = 4, and repeat the calculation, a = $3 \times 2^4 - 1$ = 47, b = $3 \times 2^3 - 1$ = 23 and c = $9 \times 2^7 - 1 = 9 \times 128 - 1 = 1151$, the pair of amicables will be $2^4 \times 47 \times 23 = 17,296$ and $2^4 \times 1151$ = 18,416, which is the same pair found by the Arab mathematician above.

Will there be another pair when n = 3? Your homework is to try n = 3 and explain why it doesn't work.

The formula works once again when n = 7. If you've got the stomach for it, you will be rewarded with 9,363,584 and 9,437,056, which are 105th on the official list.

By the way, the answer to your homework is that c = 287, which divides by 7 so is not prime.

Martin Gardner's Conjecture

It is obvious that there are far more even amicable numbers than odd ones, and in most of the odd ones the units figure is 5. In each pair, they are either both even or both odd.

Martin Gardner, the American mathematician who has done so much to popularize his subject, conjectured that the sum of every pair of amicable numbers was divisible by 9. In the pairs under 20,000,000 this seems to be the case.

But far along the number line was the enormous pair, 666,030,256 and 696,630,544. They add to 1,362,660,800. The digits add to 32, which is not on the 9 times table so, alas, there goes another promising conjecture – up in smoke.

Number Chums

$3869 = 62^2 + 05^2$ and $6205 = 38^2 + 69^2$

and

$5965 = 77^2 + 06^2$ and $7706 = 59^2 + 65^2$

But, of course, they are not amicable pairs. They are just good friends.

Pascal's Triangle

Blaise Pascal was a French mathematician, scientist and religious philosopher. He was born in 1623 and lived in poor health for only thirty-nine years. Yet in that short time, he was regarded as a brilliant man in each of his three fields of interest.

In mathematics, he is best known for his beautifully symmetrical triangle. As much as can be fitted on to the page is seen below but like so many other numerical patterns, it goes on forever.

```
                    1
                  1   1
                1   2   1
              1   3   3   1
            1   4   6   4   1
          1   5  10  10   5   1
        1   6  15  20  15   6   1
      1   7  21  35  35  21   7   1
    1   8  28  56  70  56  28   8   1
  1   9  36  84 126 126  84  36   9   1
1  10  45 120 210 252 210 120  45  10   1
1  11  55 165 330 462 462 330 165  55  11   1
```

What to notice:

1. The top 1 is not really the first row. It is there simply to give the triangle a pointed top. Sometimes, it is called Row 0.

2. Each number inside the triangle is the sum of the 2 numbers above it.

3. The actual first row, Row 1, starts 1 1, and there is a sloping row of 1s down each side.

4. The sum of the numbers in each horizontal row is a power of 2, which is the same as the row number. For example, the sum of the numbers in row 5 = 32 = 2^5.

5. The number formed by the digits in each row is a power of 11: 11^0, 11^1, 11^2, 11^3, 11^4. Ah, but 11^5 is *not* 15101051. The problem lies with the two 10s where the 1s have to be carried, making 11^5 = 161051. Likewise Row 6 is 1 6 15 20 15 6 1 so carrying the 1, 2 and 1 gives 11^6 = 1771561. OK, it gets a bit fiddly after Row 8.

6. The second sloping row with 1, 2, 3, 4, 5, 6, etc. tells you the number of each row.

7. The 3rd sloping row 1, 3, 6, 10, 15, 21, etc. is a list of the triangular numbers.

8. The 4th sloping row, 1, 4, 10, 20, 35, 56, etc. is a list of the tetrahedral numbers.

9. Look along any sloping row and do some Hockey Stick Adding. For example, come left to right down the sloping row 1, 6, 21, 56, 126 and stop and turn southwest to 210:

   1 + 6 + 21 + 56 + 126 = 210. Try it again at the other side. Come right to left down the sloping row 1, 7, 28,

84, 210 and stop and turn south-east to 330: 1 + 7 +
28 + 84 + 210 = 330.

10. Look at the numbers down the centre 'spine' of the
triangle, 1, 2, 6, 20, 70, 252 ... these are the central
binomial coefficient numbers used in algebra – oops
– a pledge was signed at the beginning not to mention
that word so let's just use your excellent grasp of
factorials to work out the next 'spine' number after
252. Get your calculator out. 252 is the sixth number
and the formula to work out the next one is:

$$S_{n+1} = (2n)! \div (n!)^2$$
$$= 12! \div (6!)^2$$
$$= 479{,}001{,}600 \div 720 \times 720$$
$$= 924$$

And if you look at the bottom line of the triangle and
add 462 and 462, you'll see that you are correct and
the seventh number in the centre is 924.

The triangle is very useful for finding the probability of events
where there are only 2 outcomes. For example, tossing a coin
will be either a head or a tail. If 3 coins are tossed, they will
land on the table in 8 different ways: HHH, HHT, HTT,
HTH, THH, TTH, THT or TTT. If you look at Row 3 of the
triangle (1, 3, 3, 1) the numbers tell us that there is only 1
chance in 8 of getting all heads or all tails but 3 chances in 8 of
getting 2 heads and a tail or 2 tails and a head.

Looking at Row 4 (1, 4, 6, 4, 1) and substituting boys and
girls for heads and tails, we see that in a family of 4 children

there can be BBBB, BBBG, BBGG, BGGG or GGGG, and the family is 6 times more likely to have 2 boys and 2 girls (in any order) than all boys or all girls.

However, apart from finding probabilities, Pascal's triangle ties up with a neat little formula for finding combinations. OK, no silly remarks from older readers, please! (Combinations, in times gone by, was the name given to a rather unattractive male undergarment covering the lower body and legs.) In maths, a combination is a way of selecting things out of a larger group *when the order does not matter.*

Suppose your friend has only 4 chocolates left in her box, but she tells you to have 3 of them and just to leave her with the last one. There is an almond cream, a butterscotch delight, a caramel fudge and a double chocolate layer – or, since you are mathematicians, A, B, C and D. So, in no particular order, you can choose ABC or ABD or BCD or ACD. In maths, what you are doing is 4C3 or 'from 4, choose 3'. Now go to Row 4 of Pascal's triangle, start at 1 and do 3 leaps to 4. So there were 4 ways you could choose your chocolates.

Now suppose your friend has 5 chocolates in the box but she invites you to choose only 2. You can have AB, AC, AD, AE, BC, BD, BE, CD, CE or DE – 10 choices. Now go to Row 5, start counting at 1 and do 2 leaps and land on 10.

One more example: your local supermarket does an all-day breakfast. For £2.50 you can choose any 3 items from fried eggs, bacon, sausages, black pudding, tomatoes, mushrooms and hash browns. How many different combinations could you have? You don't really need to bother writing them all out. But just to convince you, here they are (the 7 breakfast items are a, b, c, d, e, f, g):

abc	acd	ade	aef	afg	bcd	bde	bef	bfg	cde	cef	cfg	def	dfg	efg
abd	ace	adf	aeg		bce	bdf	beg		cdf	ceg		deg		
abe	acf	adg			bcf	bdg			cdg					
abf	acg				bcg									
abg														

Total number of ways = 35.

Here is the easy way to work this out. Go to Row 7, start at 1, do 3 leaps and land on 35. There are 35 different ways you can choose your £2.50 breakfast.

Now for the formula. It uses factorials that you learned about earlier in the 'Factorial!' section. To calculate the combination above we could have used the formula r! ÷ n!(r − n)! where r is the row number and n is the number along the row. So:

r! ÷ n! (r − n)!
= 7! ÷ 3! × (7 − 3)!
(7 × 6 × 5 × 4 × 3 × 2 × 1) ÷ (3 × 2 × 1) × (4 × 3 × 2 × 1)
= 5040 ÷ (6 × 24)
= 5040 ÷ 144
= 35

(The calculation could have been simplified by cancelling.)

Suppose we want to find a certain number in Pascal's triangle but we have only bothered to write out the first few rows. What, for example, is the 3rd number in the 9th row (counting from 9)? Using the formula r! ÷ n! (r − n)!, we have:

9! ÷ 3!(9 − 3)!
= 9! ÷ (3! × 6!)

There is no need to calculate 9! Cancel 6! into 9! and you are left with:

$$(9 \times 8 \times 7) \div (3 \times 2 \times 1) = 84$$

See the triangle above – yes, 84, is correct.

Now for the big one! What is the chance of getting all six numbers in the National Lottery?

We can find this in the forty-ninth row of Pascal's triangle, the sixth number along. However, who has the energy or width of paper to write out all these rows! So we will calculate:

$49! \div 6! \, (49 - 6)!$
$= 49! \div (6! \times 43!)$
As before, cancel 43! into 49!
$= (49 \times 48 \times 47 \times 46 \times 45 \times 44) \div (6 \times 5 \times 4 \times 3 \times 2 \times 1)$
$= 13{,}983{,}816$

So that's approximately 1 chance in 14 million.

232, 233 and 234

These ordinary-looking consecutive numbers are all the sum of 2 squares:

$$232 = 6^2 + 14^2$$
$$233 = 8^2 + 13^2$$
$$234 = 3^2 + 15^2$$

264² and Other Palindromic Squares

$264^2 = 69,696$
$836^2 = 698,896$
$2285^2 = 52,211,225$

365

Here's a neat progression of squares to make 365.

$$(10^2 + 11^2 + 12^2 + 13^2 + 14^2) \div 2 = 365$$

Days in the Year

There are 365 days, 5 hours, 48 minutes and 46.08 seconds in a year (to the nearest $\frac{1}{100}$th of a second).

512

$$512 = (5 + 1 + 2)^3$$

561 and the
Carmichael Numbers

Named in the early twentieth century by R. D. Carmichael, every Carmichael number is an odd whole number that is the product of at least 3 distinct prime numbers. The smallest one and the only one with 3 digits is 561, with its prime factors 3, 11 and 17.

But, you ask, why is 561 the smallest one? What about the following, for example?

$231 = 3 \times 7 \times 11$ or $455 = 5 \times 7 \times 13$

And many others? Well, these numbers do not obey the strict Carmichael rules. Not only do the factors of the number have to be prime but *when 1 is subtracted from each prime, they have to divide into the Carmichael number minus 1.*

In the 1st alternative suggestion, $(231 - 1)$ does not divide by $(3 - 1)$ *and* $(7 - 1)$ *and* $(11 - 1)$. In the 2nd alternative suggestion, $(455 - 1)$ does not divide by each of $(5 - 1)$, $(7 - 1)$ and $(13 - 1)$. But look at 561. 560 divides by 2, 10 and 16.

There are no other numbers under 561 that satisfy these conditions. The 2nd Carmichael number is $1105 = 5 \times 13 \times 17$. Test: 4, 12 and 16 divide 1104. The 20th Carmichael is 162,401 $= 17 \times 41 \times 233$. Test: 16, 40 and 232 divide 162,400.

Like prime numbers, Carmichael numbers grow to an enormous size but unlike prime numbers, which are infinite, it is only conjectured that the Carmichael numbers go on forever.

Similar to prime numbers and binary numbers, the Carmichael numbers were invented and studied long before they had any obvious use. Practical key encryption uses extremely long prime numbers and Carmichael numbers called 'keys'. The more digits in these keys, the more secure the process.

List of the first 8 Carmichael Numbers

$561 = 3 \times 11 \times 17$
$1105 = 5 \times 13 \times 17$
$1729 = 7 \times 13 \times 19$
$2465 = 5 \times 17 \times 29$
$2821 = 7 \times 13 \times 31$
$6601 = 7 \times 23 \times 41$
$8911 = 7 \times 19 \times 67$
$10,585 = 5 \times 29 \times 73$

593

$$593 = 23 \times 28 - 23 - 28$$

642 and 643: The Amazing Difference between 2 Cubes

$642^3 = 264,609,288$
$641^3 = 263,374,721$

So:

$642^3 - 641^3 = 1,234,567$

648 and the Other Members of the Club

648 divides exactly by the square of the sum of its digits:

$$648 \div (6 + 4 + 8)^2 = 648 \div 18^2 = 648 \div 324 = 2$$
$$972 \div (9 + 7 + 2)^2 = 972 \div 18^2 = 972 \div 324 = 3$$
$$605 \div (6 + 0 + 5)^2 = 605 \div 11^2 = 605 \div 121 = 5$$
$$392 \div (3 + 9 + 2)^2 = 392 \div 14^2 = 392 \div 196 = 2$$

The others in this exclusive club are 112, 243, 324, 405, 512 and 810.

666

666 is mentioned in Revelation, the last book of the Bible, as the Number of the Beast. It is a mystical number of unknown meaning, but the beast refers to a certain man, possibly the Roman emperor, Nero. It is interesting that 666 in Roman numerals is DCLXVI, one of each numeral.

Apart from that, 666 is full of fun and magic.

$$666 = 2^2 + 3^2 + 5^2 + 7^2 + 11^2 + 13^2 + 17^2$$

That is, the sum of the squares of the first 7 prime numbers.

$$666 = 1^6 - 2^6 + 3^6 \ (1 - 64 + 729)$$
$$666 = 1 + 2 + 3 + 4 + 567 + 89$$

And also:

$$666 = 123 + 456 + 78 + 9$$
$$666 = 6 + 6^3 + 6 + 6^3 + 6 + 6^3$$

And even going backwards:

$$666 = 9 + 87 + 6 + 543 + 21$$

And there's more!

666 is the sum of the cubes of the digits in its square plus the digits in its cube. Huh?

Let's say that slower and in bits:

$666^2 = 443{,}556$

$4^3 + 4^3 + 3^3 + 5^3 + 5^3 + 6^3$

$= 64 + 64 + 27 + 125 + 125 + 216 = 621$

$666^3 = 295{,}408{,}296$

$2 + 9 + 5 + 4 + 0 + 8 + 2 + 9 + 6 = 45$

$621 + 45 = 666$

Oh! And don't forget! 666 is also the 36th triangular number: $36 \times 37 \div 2 = 666$.

703 and the Kaprekar Numbers

D. R. Kaprekar was a twentieth-century Indian mathematician and schoolteacher. Much of his work was thought to be interesting only in recreational maths circles but lately it has been taken more seriously.

He discovered that certain numbers, when squared, split and added, result in the original number. For example, $703^2 = 494209$ and $494 + 209 = 703$.

In another example, 297, the square does not split evenly, so keep the larger number of digits to the right. $297^2 = 88209$ and $88 + 209 = 297$.

The Kaprekar numbers start 1, 9, 45, 55, 99, 297, 703 and 999 (and include every number consisting of 9 repeated, which means that they must go on forever). Here are 2 others along the number line that you may like to test: 17,344 and 538,461. But these are just 2 of the midgets. They can be very much bigger.

And, 'curiouser and curiouser' as Alice in Wonderland said, if you take any Kaprekar number K with n digits, then $K^2 - K$ will be divisible by $10^n - 1$ (numbers like 99, 999, etc.)

Let's test it for an easy one:

$$45^2 - 45 = 2025 - 45 = 1980 \div 99 = 20$$

Now, just for fun, let's take a monster: 461,539. (It really is a Kaprekar number; it's been checked.)

$$461,539^2 - 461,539 = 213,018,248,531 - 461,539 = 213,017,786,982$$

When that horrendous number is divided by 999,999, you get 213,018 exactly!

The Kaprekar numbers can be extended to higher powers. For example, $45^3 = 91125$ and $9 + 11 + 25 = 45$. Other numbers with this property are 297, 2322, 2728 and many others where the calculation is too much to bear. But here is 2728^3.

$$2728^3 = 20,301,732,352$$
$$203 + 0173 + 2352 = 2728$$

As you would have guessed, there are also 4th power and 5th power Kaprekars, most of them monsters, but 67^4 is easy to check.

$$67^4 = 20,151,121$$
$$20 + 15 + 11 + 21 = 67$$

The Kaprekar Constant 6174

Take any non-palindromic 4-digit number, for example, 3142. Arrange the digits to make the largest and smallest numbers and subtract. Repeat.

$$4321 - 1234 = 3087$$
$$8730 - 0378 = 8352$$
$$8532 - 2358 = 6174$$

OK, why not go further? Well, if you do, you'll get back to 6174 again. To convince you, we'll go on: 7641 − 1467 = 6174. Satisfied?

6174 is also one of Kaprekar's Harshad or 'joy giving' numbers (see 'Harshad Numbers' below) because it is divisible by the sum of its digits: 6174 ÷ 18 = 343.

1000

For some people, subtracting from 1000 in your head is tricky. For example, 1000 – 643. Well, 3 from 0, you cannot, borrow 1, so 3 from 10 = 7, pay back 1, so 5 from 0, you cannot, borrow 1 … oh dear!

Here's the easy way. 1000 = 999 + 1. Take 643 from 999 and pop the 1 back on again. 999 – 643 = 356 + 1 = 357.

Next time you can't get off to sleep, do the traditional thing and count sheep.

Imagine the sheepdog herding the sheep, one at a time, into the pen. Visualize each number as written-out words – one sheep, two sheep, three sheep … Exclude the joining word 'and', and keep looking for the first letter in the alphabet. Eight hundred and fifty-two … nine hundred … nine hundred and ninety-nine, one thousand. And there it is – the letter 'a' for the first time!

The Effect of Multiplying by 1001

729 × 1001 = 729,729
315 × 1001 = 315,315
504 × 1001 = 504,504

When any 3-digit number, abc, is multiplied by 1001, the product is abcabc.

A glance at 1001 shows that it is not prime like its little brother, 101. Doing any of the divisibility tests for 7, 11 and 13, you see that 1001 = 7 × 11 × 13. So that means that any 6-digit number like, for example, 218,218, will always divide by 7, 11 and 13.

218,218 ÷ 7 = 31,174
31,174 ÷ 11 = 2834
2834 ÷ 13 = 218

1033 and Other Constant Base Powers

$1033 = 8^1 + 8^0 + 8^3 + 8^3$

$4624 = 4^4 + 4^6 + 4^2 + 4^4$

$595,968 = 4^5 + 4^9 + 4^5 + 4^9 + 4^6 + 4^8$

1089

$1089 \times 1 = 1089$ and $1089 \times 9 = 9801$
$1089 \times 2 = 2178$ and $1089 \times 8 = 8712$
$1089 \times 3 = 3267$ and $1089 \times 7 = 7623$
$1089 \times 4 = 4356$ and $1089 \times 6 = 6534$
$1089 \times 5 = 5445$ but has no matching partner

A Little Number Trick Involving 1089

Take any 3-digit number that is not palindromic. Reverse it and subtract, then reverse and add. You always end up with 1089. For example:

$372 - 273 = 099$
$990 - 099 = 891$
$891 + 198 = 1089$

Also:

$1089 = 33^2$
$65^2 - 56^2 = 4225 - 3136 = 1089 = 33^2$

1138 Comes Full Circle After a 4th Power Journey

(This one deserves a gold medal.)

$1138 \neq 1^4 + 1^4 + 3^4 + 8^4$. This = 4179
$4179 \neq 4^4 + 1^4 + 7^4 + 9^4$. This = 9219
$9219 \neq 9^4 + 2^4 + 1^4 + 9^4$. This = 13,139
$13,139 \neq 1^4 + 3^4 + 1^4 + 3^4 + 9^4$. This = 6725
$6725 \neq 6^4 + 7^4 + 2^4 + 5^4$. This = 4338
$4338 \neq 4^4 + 3^4 + 3^4 + 8^4$. This = 4514
$4514 \neq 4^4 + 5^4 + 1^4 + 4^4$. This = 1138

1233 and Special Relationships

The sum of two squares is sometimes made up of the same digits:

$12^2 + 33^2 = 1233$
$88^2 + 33^2 = 8833$

The next one needs a fanfare of trumpets!

$116,788^2 + 321,168^2 = 116,788,321,168$
$(116,788^2 + 321,168^2 = 13,639,436,944 +$
$103,148,884,224 = 116,788,321,168)$

1676

$1^1 + 6^2 + 7^3 + 6^4 = 1 + 36 + 343 + 1296 = 1676$

Harshad Numbers

These numbers were defined by D. K. Kaprekar who, because he already had a special set of numbers in his own name, decided to call them Harshad numbers. 'Harshad' means 'joy giving'.

A Harshad number is an integer that is divisible by the sum of its digits – a simple enough idea. It's a wonder that the Pythagoreans didn't at least give it some consideration over 2,500 years ago.

24 is Harshad because 24 → 2 + 4 = 6 and 6 is a divisor of 24. 153 → 1 + 5 + 3 = 9, which is a divisor of 153.

The Harshad numbers are very numerous. There are 50 of them in the first 200 integers and that's not including the trivial 1-digit numbers.

6804 is said to be a multiple Harshad number (MHN) because:

$$6804 \div 18 = 378$$
$$378 \div 18 = 21$$
$$21 \div 3 = 7$$
$$7 \div 7 = 1$$

As usual, mathematicians have been searching for the biggies. Your joy will be unlimited if you test the following MHN: 2,016,502,858,579,884,466,176. You can have 108, the sum of the digits, as a free gift.

Self Numbers

Self numbers are yet another discovery of D. K. Kaprekar.

First of all, although this is a little unusual, it would be easier this time to show you what is *not* a self number. Take any number, say 35, and add on its digits, getting 35 + 3 + 5 = 43. So 43 is not a self number as it has been generated by another number, in this case 35. Take 73: 73 + 7 + 3 = 83 so 83 is not a self number because it has been generated by 73. Likewise, 19 is not a self number because 14 + 1 + 4 = 19. Some numbers, for example 101, even have two generators: 91 + 9 +1 = 101 and 100 + 1 + 0 + 0 = 101.

OK, so what *is* a self number? You might be thinking that surely all numbers are generated by some other number using this ab + a + b formula. But there are numbers that are not able to be generated in this way, for example, 20. There is no ab + a + b that adds to 20. It falls between 14 + 1 + 4 and 15 + 1 + 5.

So a number is a self number if there is no number ab + a + b or abc + a + b + c or abcd + a + b + c + d and so on to generate it.

There are only 13 self numbers under 100. The single-digit ones are 1, 3, 5, 7 and 9, and the others are 20, 31, 42, 53, 64, 75, 86 and 97, but they go on forever.

As you can see from the short list above, some self numbers are also prime numbers. Self primes start 3, 5, 7, 31, 53, 97, 211, 233, 277, 367 … and they also go on forever.

One enormous Mersenne prime number recently discovered was also found to be a self number. You want to know what this number is, don't you? Well, it is $2^{24036583} - 1$. Hats off to any reader who can even imagine how many digits are in this mega-monster self prime number.

Ramanujan and the Taxi Cab Number 1729

The story behind this number helps us remember the Indian mathematician Srinivasa Ramanujan (1887–1920). A self-taught genius, he is considered to be one of the world's greatest mathematicians, despite his poverty, lack of formal education and poor health. Lacking the money even to buy paper, he often worked with chalk on slate.

Ramanujan excelled in number theory and in 1914 he came to the attention of G. H. Hardy, an eminent English mathematician who was working in the same field. Hardy recognized his brilliance and offered to sponsor the young prodigy to study in England. They worked together until Ramanujan became terminally ill.

Not long before Ramanujan returned to India, Hardy visited his young friend in hospital. He remarked that the taxi he had used was numbered 1729, a very boring number.

Ramanujan, a mathematician to the end, disagreed. Lying on his back in the hospital bed, he was quick to point out to his mentor that 1729 was a very interesting number. It was the smallest number to be the sum of 2 cubes in 2 different ways: $1^3 + 12^3$ and $9^3 + 10^3$.

Ramanujan produced a few extraordinary approximations for π including:

$$\pi \approx (9801 \div 1103) \times (\sqrt{2} \div 4)$$

Let's check it using as many decimal places as a pocket calculator will allow:

$$\pi \approx 8.885766 \times 0.3535533$$
$$\pi \approx 3.1415918 \dots$$

Another of Ramanujan's discoveries for an approximation for π is:

$$\pi \approx \sqrt[4]{\frac{19^2}{2^2} + 9^2}$$

Don't be alarmed at the 4th root. It is easy to do without a fancy calculator. Just press the square root button twice.

Another one you can try yourself is

$$(63 \div 25) \times (17 + 15\sqrt{5}) \div (7 + 15\sqrt{5}).$$

1729 Again

1729 is also one of Kapreker's Harshad numbers as it is divisible by the sum of its own digits. $1 + 7 + 2 + 9 = 19$. It is also the 3rd Carmichael number.

And $1729 = 19 \times 91$, which is quite neat too.

Brocard's Problem

A nineteenth-century French mathematician, Henri Brocard, tried to find out which values of n would make n! + 1 a square number. It worked for n = 4. 4! + 1 = 24 + 1 = 5^2.

When n = 5, 5! + 1 = 120 + 1 = 11^2, and when n= 7, Brocard got an interesting result, 7! + 1 = 71^2 (5040 + 1).

But there were no more. Later, in 1913, Ramanujan, unaware of Brocard's efforts, tackled the same problem trying values of n up to over 60 with no calculating aids. Eventually, he became convinced there were no further solutions.

Lucky Numbers

To find the 'lucky numbers', you have to do a little work. So, if you want, get yourself a sheet of A4 paper, turn it landscape-wise and write out the natural numbers from 1 to 100:

1, 2, 3, 4, 5, 6, 7, 8, 9, 10, 11, 12, 13, 14, 15, 16, 17, 18, 19, 20, 21, 22, 23, 24, 25, 26, 27, 28, 29 …

Having done that, you are going to do a sort of Sieve of Eratosthenes on them. (Remember the third-century BC genius we talked about earlier?) But this time, it will be more of a massacre than a sieve. First eliminate *all* the even numbers, including 2. 1 remains safe forever. So you are left with:

1, 3, 5, 7, 9, 11, 13, 15, 17, 19, 21, 23, 25, 27, 29, 31, 33, 35, 37, 39, 41, 43, 45, 47, 49 …

And so on to 99.

The 2nd number not eliminated is 3, so starting at the beginning of the above row, kill off every third number starting with 5. This leaves you with:

1, 3, 7, 9, 13, 15, 19, 21, 25, 27, 31, 33, 37, 39, 43, 45, 49, 51, 55, 57, 61, 63, 67, 69, 75, 79, 81, 85, 87, 91, 93, 97, 99

The third number not crossed out is 7, so starting at the beginning of the killing field again, cross out every seventh number starting with 19. So you are left with:

1, 3, 7, 9, 13, 15, 21, 25, 27, 31, 33, 37, 43, 45, 49, 51, 55, 57, 63, 67, 69, 73, 75, 79, 81, 85, 87, 91, 93, 97, 99

Fun, isn't it? Now your next number not eliminated is 9. So start at the beginning of the survivors and kill every ninth number starting with 27, and you will be left with:

1, 3, 7, 9, 13, 15, 21, 25, 31, 33, 37, 43, 45, 49, 51, 55, 63, 67, 69, 73, 75, 79, 85, 87, 93, 97, 99

In the next slaughter, the thirteenth number, 45, will be bumped off and every thirteenth number after that. Now those left are:

1, 3, 7, 9, 13, 15, 21, 25, 31, 33, 37, 43, 49, 51, 55, 63, 67, 69, 73, 75, 79, 85, 87, 93, 99

You are now ready to slay every fifteenth number so starting at the beginning again, the first number to go is 55. It's clear that the annihilation is slowing down because 55 is the only one caught in the line of fire this time.

In the cull of the twenty-first number, only 85 falls and in the attack on every twenty-fifth number, there are no numbers left to eliminate.

So all the following numbers are safe:

1, 3, 7, 9, 13, 15, 21, 25, 31, 33, 37, 43, 49, 51, 63, 67, 69, 73, 75, 79, 87, 93, 99

You must have guessed why they are called lucky numbers. In that little battle only 39 numbers were lucky to be left standing.

Lucky Points to Notice

1. Like the prime numbers, the lucky numbers go on forever. (You tested only the first 100 integers.)
2. Comparing the lucky numbers with the prime numbers, the elimination of numbers to find primes depends on *what* they are; the elimination of the above numbers depends on *where* they are.
3. In the first 100 numbers there are 25 primes and 23 luckies. In the first 200 numbers, there are 46 primes and only 39 luckies.
4. Along the number line, there are twin primes and twin luckies, and sexy primes and sexy luckies. In the first 100 integers (7 and 13), (31 and 37) and (67 and 73) are sexy, lucky and prime. How special is that!
5. The primes go on forever, so do the luckies. But it is not certain yet if there is an infinite number of lucky primes.

The luckies feature in a conjecture, not yet proved. It's a bit like Goldbach's conjecture discussed earlier. This time it is conjectured that 'Every even number is the sum of 2 luckies.'

Like Goldbach's conjecture, it is easy to demonstrate:

$$30 = 21 + 9$$
$$44 = 37 + 7$$
$$100 = 49 + 51$$

And on and on with no exceptions to be found so far.

Good News for an Unlucky Number

At the end of the above annihilation, do you notice a particular number, near the beginning of the list of lucky numbers, which was still standing?

Yes, 13 is officially a lucky number. Isn't that good news? The most feared and maligned number for centuries has been proved by mathematics to be not guilty.

Apart from being lucky, 13 is also a prime number, an emirp and a happy number.

$$13 \rightarrow 1^2 + 3^2 = 10 \rightarrow 1^2 + 0^2 = 1$$

From now on, there's no need to exclude 13 from your lottery ticket and, if you enjoy skydiving, why not do it on the 13th of the month?

1961

This one is from a Christmas cracker. What happened in 1961 and won't happen again until 6009? Turn the book upside down and all will be revealed.

And still being silly, the following magic square works just as well upside down too.

18	99	86	61
66	81	98	19
91	16	69	88
89	68	11	96

The Untouchables

In the number world, prime numbers rule. Mathematicians love them. They are also fond of perfect numbers, like 28 where the divisors 1, 2, 4, 7 and 14 add to the number itself, and amicable number pairs like 220 and 284 where 220 is the sum of the divisors of 284 and 284 is the sum of the divisors of 220.

At the bottom of this number hierarchy are the untouchables. 88 is an untouchable number because there is no number whatsoever whose divisors add up to 88. Yes, really! The divisors of 80, for example, are 1, 2, 4, 5, 10, 16, 20 and 40, and for 84 they are 1, 2, 3, 4, 6, 7, 12, 14, 21, 28 and 42; you can see at a glance that neither example gives a total of 88. Try any other number you want but the sum of the divisors is never 88. No other number wants to have anything to do with an untouchable.

But take a number that is not untouchable, like 20, for example. The divisors of 34 are 1, 2, 17, which happily add to 20. And 21 is not untouchable either because the divisors of 18 are 1, 2, 3, 6 and 9.

The first few untouchable numbers are 2, 5, 52, 88, 96, 120, 124, 146, 162, 188, 206, 210, 216 and 238. It is conjectured that they are all even numbers except 5, and 5 and 2 are the only primes. The first few primes sometimes act out of character and you can see that the untouchables don't really get going until 52.

It won't surprise you to see that an untouchable number is never a lucky number and that includes 5.

An untouchable never stands 1 number ahead or even 3 numbers ahead of a prime. In maths notation, $p + 1 \neq u_n$ and $p + 3 \neq u_n$. Here is a short list of the relevant primes up to 167:

... 47, <u>52</u>, 53, 59, 61 ... 83, <u>88</u>, 89, <u>96</u>, 97, 101 ... 157, <u>162</u>, 163, 167 ...

You can see that, although the untouchable never appears 1 or 3 numbers *ahead* of a prime, a sneaky underlined untouchable is often very close *behind*.

It has been proved that the untouchable numbers go on forever.

2187 and Family Members

2187 is a lucky number further along the number line, one that Martin Gardner was fond of simply because it happened to be the door number of his boyhood home. It also equals 3^7. And 2187 and 7812 = 9999.

Take the same digits in another order, 8127, and multiply 81 by 27 and you get 2187. Split these digits and multiply: 21 × 87 = 1827.

Take another arrangement of the digits: 2178 × 4 = 8712, its own reversal.

And look at 2178: $2^4 + 1^4 + 7^4 + 8^4 = 6514$, but $6^4 + 5^4 + 1^4 + 4^4 = 2178$.

Write down all the 2-digit numbers you can make with 1, 7, 8 and 2. Add them and multiply by 3.

(17 + 18 + 12 + 78 + 72 + 82 + 71 + 81 + 21 + 87 + 27 + 28) × 3 = 594 × 3 = 1782

One more arrangement of the digits: 1728 is the number of cubic inches in a cubic foot and if we ever do adopt a duodecimal system as suggested earlier in the book, $1728_{10} = 1000_{12}$.

2519

$$1259 \times 2 + 1 = 2519$$
$$839 \times 3 + 2 = 2519$$
$$629 \times 4 + 3 = 2519$$
$$503 \times 5 + 4 = 2519$$
$$419 \times 6 + 5 = 2519$$
$$359 \times 7 + 6 = 2519$$
$$314 \times 8 + 7 = 2519$$
$$279 \times 9 + 8 = 2519$$
$$251 \times 10 + 9 = 2519$$

2592

$2592 = 2^5 \times 9^2$ (a nice Friedman number)

The Amazing 2880

$721^2 - 719^2 = 2880$

And so do all these others.

$362^2 - 358^2$
$243^2 - 237^2$
$184^2 - 176^2$
$149^2 - 139^2$
$126^2 - 114^2$
$98^2 - 82^2$
$89^2 - 71^2$
$82^2 - 62^2$
$72^2 - 48^2$
$63^2 - 33^2$
$61^2 - 29^2$
$58^2 - 22^2$
$56^2 - 16^2$
$54^2 - 6^2$

The quick way to check all of them is by using the difference of 2 squares formula, $X^2 - Y^2 = (X - Y)(X + Y)$. For example

$$149^2 - 139^2 = (149 - 139)(149 + 139) = 10 \times 288 = 2880$$

You're not finished with 2880 yet!

Start with 482 and add 2880 and then add another 2880 as follows:

$$482 + 2880 = 3362$$
$$3362 + 2880 = 6242$$

So we have 3 different numbers: 482, 3362 and 6242. Add them in 3 different ways,

$$482 + 3362 = 3844 = 62^2$$
$$482 + 6242 = 6724 = 82^2$$
$$3362 + 6242 = 9604 = 98^2$$

3367

3367 can be multiplied in your head by any 2-digit number. For example, 3367 × 69 = 232323 and 3367 × 36 = 121212. You don't need to be an arithmetical genius. Of course, it is easier when the multiplier is divisible by 3 but 3367 × 28 isn't that difficult either: 282828 ÷ 3 = 94276.

Here's how it works. Remember 101 and its palindromic pals? 10101 = 3 × 3367 and any 2-digit number ab multiplied by 10101 is ababab, so 3367 × 45 = 454545 ÷ 3 = 151515.

3435 and Münchhausen

Baron Hieronymus von Münchhausen was an eighteenth-century German nobleman. He was notorious for being the raconteur of tall tales about his escapades, including a trip to the moon, in order to make himself more interesting and popular.

3435 is called a Münchhausen number because it is about as interesting as any number can be. 3435 is equal to the sum of its digits raised to each digit's power:

$$3^3 + 4^4 + 3^3 + 5^5 = 27 + 256 + 27 + 3125 = 3435$$

Now who could deny that 3435 is pretty special? It is likely that it is the only one in the whole number line.

But there is another number even more spectacular than 3435. It is 438,579,088. However, $4^4 + 3^3 + 8^8 + 5^5 + 7^7 + 9^9 + 0^0 + 8^8 + 8^8$ has a flaw. Do you see it? It's the zero bit. Mathematicians are still debating whether $0^0 = 0$ or $0^0 = 1$ or it may forever be 'undefined'. There is no need to calculate each term. Without the 0^0, the other terms will be even, odd, even, odd, odd, odd, even and even, and the total would be even. So 0^0 needs to be 0 to make the final total an even number.

If it is 0, then, surely, it is the most famous number ever discovered and Münchhausen would be delighted to claim it. But, sadly, if $0^0 = 1$, well, that 9-digit number is one of the greatest disappointments in mathematics.

The other name for these numbers is perfect digit-to-digit invariants or PDDIs but Münchhausen is more memorable, isn't it?

A Reverse Münchhausen Number

$$48,625 = 4^5 + 8^2 + 6^6 + 2^8 + 5^4$$

4884, Palindrome by Reversal

87 + 78 = 165
165 + 561 = 726
726 + 627 = 1353
1353 + 3531 = 4884

And since 4884 is palindromic, you can go no further.

4913

4913 is the smallest 4-digit number that is the cube of the sum of its digits.

$$(4 + 9 + 1 + 3)^3 = 17^3 = 4913$$

Another example is $5832 = (5 + 8 + 3 + 2)^3$.

5777 and 5993
(2 Flies in the Ointment)

Goldbach's famous conjecture (see 'Goldbach's Conjecture') proposing that every even number greater than 2 is the sum of 2 primes has never been proved, although no counter-example has ever been found.

However, in 1752, Goldbach sent a letter to his friend, Euler, asking him to look over another conjecture. He was proposing this time that every odd number that is *not* prime could be written as a prime plus twice a square:

$p + 2a^2 \ (a > 0)$

He had obviously noticed from the start that it did not work for the primes 3 and 17. But it seemed to work for composite odd numbers:

$9 = 7 + 2 \times 1^2$
$15 = 7 + 2 \times 2^2$
$21 = 3 + 2 \times 3^2$
$25 = 7 + 2 \times 3^2$
$27 = 19 + 2 \times 2^2$
$33 = 31 + 2 \times 1^2$

And so on.

Euler replied that the conjecture looked very promising indeed.

However, a hundred years later, in 1856, Professor Moritz Stern of the University of Göttingen became interested in the conjecture and found that the composite numbers 5777 (53 × 109) and 5993 (13 × 461) did not work.

Fortunately, Goldbach was long dead before these two very disappointing counter-examples were discovered.

Goldbach may not have included prime numbers in his conjecture, having noticed that 3 and 17 failed, but it might have been a small compensation for him to learn that when over 10 million odd numbers, both prime and composite, were tested by computer, most prime numbers also satisfy the $p + 2a^2$ formula. Only 5 others failed: 137, 227, 977, 1187 and 1493.

9109

The prime number 9109 has a pretty pattern when multiplied by 1 to 9 and the digits of each product are added.

$$9109 \times 1 = 09109 \rightarrow 19$$
$$9109 \times 2 = 18218 \rightarrow 20$$
$$9109 \times 3 = 27327 \rightarrow 21$$
$$9109 \times 4 = 36436 \rightarrow 22$$
$$9109 \times 5 = 45545 \rightarrow 23$$
$$9109 \times 6 = 54654 \rightarrow 24$$
$$9109 \times 7 = 63763 \rightarrow 25$$
$$9109 \times 8 = 72872 \rightarrow 26$$
$$9109 \times 9 = 81981 \rightarrow 27$$

We can also admire the columns, which run from 1 to 9 and back again.

12,496 and the Sociable Set

12,496 is the first of a set of sociable numbers that show off their divisors to each other. As in the perfect numbers, the divisors include 1 but not the number itself.

Divisors of 12,496	Divisors of 14,288	Divisors of 15,472	Divisors of 14,536	Divisors of 14,264
1	1	1	1	1
2	2	2	2	2
4	4	4	4	4
8	8	8	8	8
11	16	16	23	1783
16	19	967	46	3655
22	38	1934	79	7132
44	47	3868	92	
71	76	7736	158	
88	94		184	
142	152		316	
176	188		632	
284	304		1817	
568	376		3634	

Divisors of 12,496	Divisors of 14,288	Divisors of 15,472	Divisors of 14,536	Divisors of 14,264
781	752		7268	
1136	893			
1562	1786			
3124	3572			
6248	7144			
Total = 14,288	Total = 15,472	Total = 14,536	Total = 14,264	Total = 12,496

So, it's back to 12,496 again.

This is the only known manageable set of sociable numbers. It was discovered by a Belgian mathematician, Paul Poulet, at the beginning of the twentieth century. Another set starts at 14,316 but it needs 28 links before it completes the chain.

We don't know how Poulet discovered 12,496. Are there any other numbers that might work *eventually*? Some might, of course, come up against a prime number, which would be interesting – but not sociable as the chain would be broken.

19,937: A Circular Prime

A circular prime is one that remains prime when, in turn, each digit is moved to the right.

19937
99371
93719
37199
71993

27,594

27,594 can be calculated in 2 amazing ways:

27,594 = 73 × 9 × 42
27,594 = 7 × 3942

40,585

40,585 = 4! + 0! + 5! + 8! + 5! (remember that 0! = 1)

Easy Adding

This is a good number trick to do when you are on a long train journey and the children in the family need a little diversion.

Ask child A to write down any 5-digit number. Then you (Y) write your 5-digit number underneath. This is repeated with child B and Granddad (G), who is reading his book, volunteers a final number.

So it goes like this:

A: 1 5 3 4 7
Y: 8 4 6 5 2
B: 2 3 7 9 8
Y: 7 6 2 0 1
G: 5 8 4 2 4

Then, here's the bad bit. You ask A or B to add it up. They grumble, 'We don't do sums on holiday.' So with a casual glance at the figures, you say, 'That's easy, it's 258422.'

Of course, they can't wait to prove you wrong.

The secret: they won't have noticed that all your digits are '9' complements (they write 2, you write 7, they write 5, you write 4). Granddad's number is totally random but it is the only important one. You attach 2 to the beginning of his number and subtract 2 from the last digit, getting 258422.

You can adapt this trick for more or fewer digits, and add other victims. The secret is all to do with the fact that 9999 + 9999 = 19,998, which is 2 short of 20,000 (or 999 + 999 + 999 = 2997, which is 3 short of 3000). Get it?

142,857

The reciprocal of 7 is 1 ÷ 7 = 0.14285714285714285714285771 42857 ... After 6 digits, it repeats itself forever.

Let's take one chunk. The first thing you may notice is that 142 + 857 = 999 and 14 + 28 + 57 = 99 and 142857 divides by 9.

Now look at the following multiplications. See how each answer is 142857 moved clockwise a little bit.

```
142857 × 1 = 142857
142857 × 2 = 285714
142857 × 3 = 428571
142857 × 4 = 571428
142857 × 5 = 714285
142857 × 6 = 857142
```

Admire the little family – never a digit out of place. And notice that not only do the rows add up to 27 but so do the columns.

When 142857 is multiplied by 7, it goes mad. 999999, it shouts, and after that the pattern seems to disappear. But does it? 142857 × 8 = 1142856 – a bit like the above pattern, isn't it? – but not good enough. Try splitting the number into 1 and 142856 and add them, and you're back to 142857 again. It works for almost any number. Try 142857 × 326. The product this time is 46571382 – nothing like the magic number. But, split it into 46 and 571382 and add. There you are, 571428, one of the family above.

However, 142857 didn't work with 7 and in fact the number doesn't like any of the multiples of 7. 142857 × 35 = 4999995 but split the number into 4 + 999995 and add, and what do you get? Yes, all the 9s again.

Try it once again for 142857 × 8764. 8764 must be a multiple of 7 because the answer is 1251998748, but split the number into 1251 and 998748, and 999999 appears again.

142,857 in Bits

$1 \times 7 + 3 = 10$
$14 \times 7 + 2 = 100$
$142 \times 7 + 6 = 1000$
$1428 \times 7 + 4 = 10,000$
$14285 \times 7 + 5 = 100,000$
$142857 \times 7 + 1 = 1,000,000$

Since 142857 repeats, you can go on and on. For example, $142857142 \times 7 + 5 = 999,999,999$. And the icing on the cake! It's a Kaprekar number. $142857^2 = 20,408,122,449$, and if you split it into 20408 + 122449 you get – would you believe it! – 142,857.

147,852

This number is not a circular version of 142,857. The digits are the same but the 7 and 2 have changed places. However, there is still some magic hanging around these digits because:

$147,852 = 333 \times 444$

And reversing the digits:

$258741 = 333 \times 777$

221,859

$221,859 = 22^3 + 18^3 + 59^3$

510,510

$510,510 = 2 \times 3 \times 5 \times 7 \times 11 \times 13 \times 17$

This is the product of the first 7 prime numbers.

$510,510 = 13 \times 21 \times 34 \times 55$

This is the product of 4 Fibonacci numbers.

2,646,798

$2,646,798 = 2^1 + 6^2 + 4^3 + 6^4 + 7^5 + 9^6 + 8^7$

Which is just about checkable without a calculator.

87,539,318 – Another Taxi Cab Number

Since Ramanujan's first taxi cab number in 1920 (see 'Ramanujan and the Taxi Cab Number 1729'), mathematicians have been hunting for more of them.

This one is Ta_3 = 87,538,318, which is the smallest number that can be expressed as the sum of 2 cubes in 3 different ways.

87,539,319
$= 167^3 + 436^3$
$= 228^3 + 423^3$
$= 255^3 + 414^3$

Interesting 8-Digit Numbers

Take any 4-digit number like 5483 and repeat it, getting the 8-digit number 54,835,483. Get out your calculator and divide this number by 73 and 137.

$$54,835,483 \div 73 \div 137 = 5483$$

Similarly:

$$12,431,243 \div 73 \div 137 = 1243$$

You want to know why, don't you? Well, $73 \times 137 = 10001$. So any 4-digit number abcd × 10001 equals abcdabcd.

123,456,789

123,456,789 × 8 = 987,654,312

(Not quite perfect, but memorable all the same.)

A Little More of 123,456,789

123,456,789 is not prime. Add the digits in the Gauss way (see the 'Karl Friedrich Gauss' section), $9 \times 10 \div 2 = 45$, and you'll see it divides by 3. But chop off the 1 at the beginning of the number and it becomes a prime.

'Miracle' Numbers

'God made the whole numbers. Everything else is the work of man.' So said the nineteenth-century Prussian mathematician, Leopold Kronecker. Having worked your way through this book of numbers, you may be inclined to agree with him. Here are a few of the author's favourite 'God made' numbers:

a) 142,857. Already discussed and the author's 'top of the pops'. Schoolchildren find it fascinating.
b) The reciprocal of 19. It was displayed on the author's classroom wall for thirty-six years.
c) 666,666 square units, the area of the right-angled triangle with sides, 693, 1924 and 2045. You can probably understand now why Pythagoras worshipped numbers.
d) 200597, a prime number and the date of birth of the author's twin granddaughters.
e) 13, everyone's new favourite number –a lucky, happy, emirpy prime.

One Billion
(1,000,000,000)

Our national debt is measured in billions of pounds and that is depressing. If something cataclysmic happens to the earth, the nearest habitable planet, if there is one, will be billions of miles away. That is unthinkable.

But what about time? What is a billion seconds?

1,000,000,000 seconds
$\approx$ 16,666,667 minutes
$\approx$ 277,778 hours
$\approx$ 11,574 days
$\approx$ 32 years

Going back from the year 2013, that's 1981. Hey, that's not so long ago! Computer games were just appearing but most teenagers were enjoying video games, *Raiders of the Lost Ark* and *Star Wars*.

What about a billion minutes? Missing out one of the divisions by 60 gives us 1903 years ago or about AD 110. Hurrah! Nicomachus may have still been alive then.

A billion hours ago, our ancestors were living in the Stone Age and the most brilliant mathematicians were the ones who could count 'One rock, two rocks ... er ... many rocks.'

1,000,000,007

Lastly, this is the smallest 10-digit prime with the maximum number of repeated digits.

And that's it! It's time to stop now before all these zeros whizz into infinity.

Postscript

Postscript

G. H. Hardy said, 'The mathematician's patterns, like the painter's or the poet's, must be beautiful; the ideas, like the colours or the words, must fit together in a harmonious way. Beauty is the first test: there is no permanent place in this world for ugly mathematics.'

Sir Andrew Wiles said, 'A discovery in mathematics is a bit like discovering oil. But mathematics has one great advantage over oil. No one has yet found a way that you can keep using the same oil forever.'

Glossary

Cube See 'Powers'.

Digit The digits that make up the number four hundred and two are 4, 0, and 2.

Divisor A number that divides into another number exactly. The divisors of 15 are 1, 3, 5, 15. Often, 15 is omitted.

Factor Another word for 'Divisor'.

Integer All whole numbers, whether positive, negative or zero, are integers.

Natural numbers The counting numbers 1, 2, 3 and so on are sometimes called *natural numbers* to distinguish them from 'Whole numbers', which include zero.

Powers The product obtained when a number is multiplied by itself a certain number of times.
5^2 (5 squared) = 5×5 = 25.
10^3 (10 cubed) = $10 \times 10 \times 10$ = 1000.
2^5 (2 to the power of 5) = $2 \times 2 \times 2 \times 2 \times 2$ = 32.
1^9 (1 to the power of 9) = $1 \times 1 \times 1 \times 1 \times 1 \times 1 \times 1 \times 1 \times 1$ = 1 (1 to any power = 1).
But *every number* (with one exception) to the power of zero is 1. For example, 1^0 = 1; 9^0 = 1.

Product   A number obtained by multiplying. The product of 2 and 3 is 6.

Reciprocal   A number turned upside down so it is '1 over'. The reciprocal of 2 is 1 over 2, which is ½, and the reciprocal of 4 is ¼. The decimal reciprocals of these are 0.5 and 0.25.

Square   See 'Powers'.

Square Root   The symbol is √. √9 = 3 because 3 multiplied by itself is 9.

Whole numbers   These start 0, 1, 2, 3 and go on forever. Whole numbers can be defined to exclude zero, and thus be the same as 'Natural numbers', or to include negative whole numbers, and thus to be the same as 'Integers'.

<   means 'is less than'.

\>   means 'is more than'.

≈   means 'is approximately equal to'.